Small Worlds

PRINCETON STUDIES IN COMPLEXITY

TITLES IN THE SERIES

Lars-Erik Cederman, *Emergent Actors in World Politics: How States and Nations Develop and Dissolve*

Robert Axelrod, *The Complexity of Cooperation: Agent-Based Models of Competition and Collaboration*

Peter S. Albin, *Barriers and Bounds to Rationality: Essays on Economic Complexity and Dynamics in Interactive Systems.* Edited and with an introduction by Duncan K. Foley

Duncan J. Watts, *Small Worlds: The Dynamics of Networks between Order and Randomness*

Scott Camazine, Jean-Louis Deneubourg, Nigel R. Franks, James Sneyd, Guy Theraulaz, Eric Bonabeau, *Self-Organization in Biological Systems*

Peter Turchin, *Historical Dynamics: Why States Rise and Fall*

Small Worlds

The Dynamics of Networks between Order and Randomness

DUNCAN J. WATTS

PRINCETON UNIVERSITY PRESS
PRINCETON AND OXFORD

Published by Princeton University Press, 41 William Street,
Princeton, New Jersey 08540
In the United Kingdom: Princeton University Press,
3 Market Place, Woodstock,
Oxfordshire, OX20 1SY

Eighth printing, and first paperback printing, 2004
Paperback ISBN 0-691-11704-7

The Library of Congress has cataloged the cloth edition of this book as follows

Watts, Duncan J., 1971–
Small worlds : the dynamics of networks between order and randomness / Duncan J. Watts.
p. cm. — (Princeton studies in complexity)
Includes bibliographical references and index.
ISBN 0-691-00541-9 (alk. paper)
1. Graph theory. 2. Network analysis (Planning). 3. Social networks—Mathematical models. I. Title. II. Series.
QA166.W38 1999
511'.5—dc21 98-56088

British Library Cataloging-in-Publication Data is available

This book has been composed in Times Roman

Printed on acid-free paper. ♾

www.pupress.princeton.edu

Printed in United States of America

10 12 14 16 15 13 11

ISBN-13: 978-0-691-11704-1

ISBN-10: 0-691-11704-7

Dedicated to the memory of David D. Browne
(1969–1998)

A man of many friends

The earth to be spann'd, connected by network,
The races, neighbors, to marry and be given in marriage,
The oceans to be cross'd, the distant brought near,
The lands to be welded together

Walt Whitman, *Passage to India*

Contents

Preface

I will be brief. Not nearly as brief as Salvador Dali, who gave the World's shortest speech. He said, "I will be so brief I have already finished," and he sat down.

E. O. Wilson, Commencement address at Penn State

This project began almost three years ago, in the middle of another miserable Ithaca winter, when I walked into Steve Strogatz's office with a vague proposition and not much hope. At the time I was doing my Ph.D. dissertation on the synchronisation of biological oscillators—crickets to be precise—and I had encountered something of a problem. The nightly struggle to attract a mate has turned the males of this particular species of cricket into highly tuned synchronising machines. Each cricket listens for the chirps of its brethren and then adjusts its subsequent chirps in response, such that the entire population winds up "oscillating" as one. But the question in my mind was Who was listening to whom? One could always assume that, within a given tree, every cricket was listening to every other. On the other hand, maybe they paid attention to only a single, nearby competitor. Or maybe it was some complicated combination in between. In any case, did it even matter how they were "connected," or would the population, as a whole, do roughly the same thing regardless? I certainly didn't know, and neither, it seemed, did anybody else. At about the same time, I was also thinking about something my father had once told me: "Did you know," he said, "that you are ever only six handshakes away from the President of the United States?" I didn't, but it seemed like an intriguing idea, and I wondered how one would go about proving it. These two ideas—one to do with the dynamics of coupled-oscillator systems, and the other concerning what I later understood to be the small-world phenomenon—eventually came together in my head, but more in the form of a daydream than a research project.

Nevertheless, this is precisely the idea that I pitched to my advisor that November day: Why don't we try to figure out what it is about the world that puts us all within a few handshakes of each other and what that implies about the dynamics of systems that are like that?

I knew that, if this idea ever turned out to be worthwhile, it would span several different disciplines, most of which neither I nor Steve knew very much about. I also knew that I wanted to graduate in less than two year's time. So when I proposed my idea with a mix of used-car-salesman enthusiasm and frightened-schoolboy stammering, I fully expected to be laughed out of his office or at least patted on the head and told to worry about something more feasible. Instead, to my endless amazement, Steve not only liked it, but he agreed to let me work on it: "We're both smart guys . . . we'll figure something out." And so we did, both of us learning and teaching in what became more of an adventure than a dissertation—the kind of research that I had always imagined research was *supposed* to be. But without that meeting and without that particular advisor, the journey would have finished before it ever had begun. I owe Steve Strogatz a lot, for many reasons, but for that first step most of all.

Of course any real journey, even one undertaken alone, succeeds only by the good grace and assistance of generous souls who have nothing to gain other than the satisfaction of having lent a hand. Foremost amongst these are Richard Rand and Dave Delchamps, for sticking with me over the course of two advisors and three vastly different research projects: their tenacity is a tribute as much to their patience as it is to their broad-mindedness. I am also deeply grateful to Frank Moon for his generosity and tolerance over the years and to Bill Holmes, upon whose assistance I have called far more often than is reasonable. For their valuable contributions to specific parts of the project, I am indebted to Dave Shmoys and Mark Huber, Brett Tjaden (alias The Oracle of Bacon), Jim Thorp and Koeunyi Bae, Martin Chalfie, Joel Cohen, Nick Trefethen, and Melanie Mitchell. Kevin MacEwen deserves a special mention for listening to me complain so much and for helping me out when I needed it most. Not enough nice things can be said about my family and friends, without whose gentle barbs and undisguised amusement I would be stuck contemplating the abstract even more than I do now. Speaking of which, a big thanks to Harrison White and also to the Santa Fe Institute, whose generous support has enabled me to continue studying the small world without having to join the real one.

Finally, I should say something about the content and organisation of this book. Since starting this work, I have been exposed to multiple literatures—from theoretical computer science to sociology to neurobiology—in which networks are important, and recognised as such. Nevertheless, the systematic analysis of networks that lie somewhere in the messy regime between order and randomness seems to be a genuinely new (or at least poorly explored) problem. Even less well understood is the relationship between the structural properties of such networks and the kind of dynamics that they lead to. How do cascading failures propagate through large power grids? Or financial systems? What is the most efficient and robust architecture for an organisation? Or a massively parallel computer? These are not just open questions—they are representative of whole fields of open questions that beckon for attention. This book is one person's attempt to define where those questions lie, and to make some microbial progress toward answering them. As such it is written not in the retrospective sense of a review—because there is not yet an established subject to be reviewed—but in the exploratory manner in which the research itself was done.

By the time this book is actually printed, some of what is contained herein will already be in need of revision, and multiple additions and refinements will no doubt have been made. Indeed, since the first published version of this work appeared a few months ago in the journal *Nature* (Watts and Strogatz 1998), Steve Strogatz and I have been contacted by dozens of researchers from virtually every discipline (yes, including English literature), who have all been struck by the relevance of small-world networks to *something* in their field of expertise. It is still too early to chart all this fascinating new terrain in detail, so treat this book less as a map than as a signpost—one that I hope might stimulate you to do some exploring yourself.

Santa Fe, New Mexico
October 1998

D.J.W.

Small Worlds

1

Kevin Bacon, the Small World, and Why It All Matters

The Oracle Says:
Presley, Elvis has a Bacon number of 2:
Presley, Elvis was in *King Creole* (1958) with Matthau, Walter
Matthau, Walter was in *JFK* (1991) with Bacon, Kevin

Excerpt from The Oracle of Bacon, A web site devoted to the Kevin Bacon Game, by Brett Tjaden and Glenn Wasson

The Kevin Bacon Game is a curious thing to be sure. For those who don't know him, Kevin Bacon is an actor best known for *not* being the star of many films. But a few years ago, Brett Tjaden—a computer scientist at the University of Virginia—catapulted Bacon to true international recognition with the claim that he was somehow at the centre of the movie universe. This is how the game goes:

- Think of an actor or actress.
- If they have ever been in a film with Kevin Bacon, then they have a "Bacon Number" of one.
- If they have never been in a film with Kevin Bacon but have been in a film with somebody else who *has*, then they have a Bacon Number of two, and so on.

The claim is that no one who has ever been in an American film *ever* has a Bacon Number of greater than four. Elvis Presley, for example, has a Bacon Number of two; his "path" is traced at the beginning of this chapter. For real enthusiasts, Tjaden created a web site that provides the Bacon Number and shortest path to the great man for the most obscure of choices.[1] In fact, Tjaden later fireproofed his claim by conducting an exhaustive survey of the Internet Movie Database,[2] which references almost all movies ever made, and determined that the highest finite Bacon Number (for *any* nationality) is eight.[3] This may seem like nothing more than another quirky fact about an already bizarre industry, but in fact it is a particularly clear example of a phenomenon that increasingly pervades our day-to-day existence: something known as the *small-world phenomenon*.

The small-world phenomenon formalises the anecdotal notion that "you are only ever six 'degrees of separation' away from anybody else on the planet." Almost everyone is familiar with the sensation of running into a complete stranger at a party or in some public arena and, after a short conversation, discovering that they know somebody unexpected in common. "Well, it's a small world!" they exclaim. The small-world phenomenon is a generalised version of this experience, the claim being that even when two people do *not* have a friend in common, they are separated by only a short chain of intermediaries. Stanley Milgram (1967) made the first experimental assault on the problem (confined to the United States) by sending a series of traceable letters from originating points in Kansas and Nebraska to one of two destinations in Boston. The letters could be sent only to someone whom the current holder knew by first name and who was presumably more likely than the holder to know the person to whom the letter was ultimately addressed. By requiring each intermediary to report their receipt of the letter, Milgram kept track of both letters and the demographic characteristics of their handlers. His results indicated a median chain length of about six, thus supporting the notion of "six degrees of separation," after which both a play (Guare 1990) and the movie adaptation have since been named. This result was both striking and surprising and continues to be so today, because the conscious construction of such chains of intermediaries is very difficult to do. Ordinarily, our perception of the social world is confined to our group of immediate acquaintances, and within this group there is a great deal of redundancy; that is, within any one circle of acquaintances, most of them know each other. Furthermore, our average *number* of acquaintances is very much less than the size of the global population (at most thousands, compared with billions). So the claim that some very short chain of acquaintances exists that links us to any other person, anywhere in the world, does seem unlikely. Since Milgram's famous experiment, much theoretical and empirical work has been performed to determine, for various social groupings:

- The characteristic number of "handshakes" between members,
- The expected number of "friends" that each member has, and
- The structure of the group, which relates one member's "circle of friends" to those of other members.

Interesting, and often creative, though this work has been (see Kochen 1989a for an excellent review), it has suffered from a number of methodological and phenomenological difficulties:

1. Detailed data of "who knows whom" is extremely hard to come by for sufficiently large groups.
2. People are notoriously bad at estimating the number of "friends" they possess.
3. Some friendships are more important than others, and some people are vastly more significant than others in connecting a network (for example, Kevin Bacon owes much of his eminent connectibility to his appearances with superstars like Jack Nicholson and Robert DeNiro).
4. Friendships are not symmetric: that is, subordinates are more likely to regard themselves as connected to their superiors than vice versa.
5. The whole notion of "friendship" is highly dependent upon both the social context (Amish farmers in rural Pennsylvania probably have quite different views of what a friendship necessitates than do Hollywood movie stars) and the nature of the question being asked (that is, friendship for the purpose of borrowing money is quite different from friendship for the purpose of spreading rumours).

Hence any conclusion drawn from such studies is susceptible to the claim that it hangs on any number of arbitrary assumptions and so does not necessarily indicate much about the world in general. Perhaps a stronger result could be obtained if we were to alter the question from Do we actually live in a small world? to What are the most general conditions under which the world can be "small"? Part I of this book is devoted to tackling this question.

Chapter 2 is a brief review of what is already understood about the small-world phenomenon, followed by a crash course in the theory of graphs—a useful and precise language that naturally lends itself to the discussion of networks. The difficulties encountered by previous researchers help motivate and contextualise the current work. Basically, the intent is to answer the reformulated question above by considering a broad range of network structures and identifying where, if anywhere, in this family of possible "worlds" dramatic changes in global network characteristics occur.

This question is approached in Chapter 3 through the introduction of two classes of graph-theoretic models: *relational graphs* and *spatial graphs*, which are distinguished fundamentally by the fact that spatial graphs contain an *external length scale* and relational graphs do not. Nevertheless, both classes consist of a one-parameter family of graphs that interpolates, as a function of the parameter, between ordered, lattice-like graphs and random graphs. Within each of these broad classes, a number of models can be constructed that turn out to be equivalent. Within the relational-graph class, two models are examined: an α-model, which is motivated by consideration of real social networks and how people actually make new acquaintances, and a β-model, which is motivated by simplicity of construction and the clean interpretation of its limiting cases. Despite apparent differences, both in motivation and construction, these two models exhibit an underlying structural similarity that can be captured by the idea of *random rewiring* and that allows their statistical properties to be expressed in a model-independent fashion. This motivates the construction of a third model, which embodies the random rewiring concept explicitly and which unifies the properties of the α- and β-models. The main result is the identification of a class of graphs—*small-world graphs*—that appear to embody the defining characteristics of the small-world phenomenon. A similar analysis is performed for spatial graphs, whose defining probability distributions have (effectively, for small n and k) a finite cutoff, but it turns out that the properties of spatial and relational graphs cannot be reconciled.

Based on the intuition gained from these simulations, a heuristic construction is posed in Chapter 4 that yields analytic approximations for the relevant statistics of both spatial and relational graphs. These approximations exhibit a surprisingly close fit to the numerical data, within their range of validity, and also predict the same scaling features as those inferred from the numerical models. As a result, the structural dissimilarity between relational and spatial graphs can be explained.

Finally, Chapter 5 deals with the length and clustering properties of three real networks, for which complete descriptions are available. All three turn out to exhibit the small-world phenomenon, leading to the speculation that this sort of property will be found to be widespread amongst many naturally occurring networks.

Part II is then an attempt to answer the equally pressing question Why should anybody care about the small-world phenomenon? Part I demonstrates that a relatively tiny amount of random rewiring can cause a dramatic reduction in the characteristic path length of a graph. But can a correspondingly minor change in the coupling topology of a distributed dynamical system have a correspondingly large effect on the *dynamical* behaviour of the system? Of particular interest are *small-world dynamical systems* (dynamical systems coupled by small-world graphs) and whether or not they exhibit any interesting properties. A series of qualitatively distinct, coupled dynamical systems are examined roughly in order of increasing complexity. Chapter 6 deals with a simple model of the spread of an infectious disease through a structured population, and Chapter 7 examines the impact that random rewiring has upon the global computational capacity of a cellular automata. Chapter 8 is a tentative exploration of the emergence and evolution of cooperation on graphs, using the iterated, multiplayer Prisoner's Dilemma. Finally, in Chapter 9 the scope is broadened to include continuous dynamical systems, examining the conditions under which systems of coupled phase oscillators can spontaneously lock into macroscopic, mutually entrained clusters. In each of these applications, the system is described, as well as why it is of interest and what is already understood about its behaviour in more familiar coupling topologies. Numerical simulations of the system are then generated for a range of graph topologies, using the algorithms of Chapter 3. The results sometimes exhibit strong correlations to the structural characteristics of the underlying graph, and sometimes they don't. A number of interesting observations can be made but no general general conclusions, other than that it is a hard and interesting problem.

So what are some possible directions in which small-world research might head? Of course, sociologists have already conceived of many applications of small world ideas (see Part 1 of Kochen 1989a Kochen 1989a for some specific examples), but these revolve principally around pure networking issues. Such applications are important for problems concerned with devising more effective schemes of marketing products or ideas, exerting influence in organisations, defining the representativeness of political bodies, and even hunting for a job. These are important ideas to be sure and appropriate for sociological research, but the interest here is focused primarily on how systems *behave* and how that behaviour is affected by their connectivity. The applications of this kind of research both encompass and stretch well beyond sociological problems, including:

1. The emergence and evolution of cooperative behaviours in large organisations, whose structural nature is allowed to vary.
2. The spread of everything from computer viruses to infectious or sexually transmitted diseases.
3. The processing of information in spatially extended and irregularly connected networks such as the human brain.

4. The design of power and communications networks (such as the Internet or cellular phone networks) to ensure rapid and cost-efficient transmission without sacrificing robustness.
5. The emergence of global computational capability from locally connected systems.
6. The synchronisation of biological oscillators, such as neurons in the brain.
7. The train of thought followed in a conversation or succession of ideas leading to a scientific breakthrough. (Here one might imagine the connections between ideas in "idea space" or conversation topics in "topic space" as one idea or topic *leading* to another that is more or less closely related.)
8. New theories of market economics that account for the network nature of transactions.
9. Improved search algorithms for optimal strategies to complex problems.
10. The formation and spread of fame, fads, and social movements.
11. A rigorous proof that Kevin Bacon really is the centre of the civilised world, and, if not, who is? (See Chapter 5 for the answer.)

The beginnings of some of these applications are addressed here and also (in dramatically condensed form) in a paper with Steven Strogatz (Watts and Strogatz 1998), but most of this work is left for future challenge. The results are far from clear, but it does seem reasonable to expect that an improved understanding of the small-world phenomenon and its generalisation to the interaction of network structure and dynamics is an important research programme with widespread applications amongst the social, mathematical, and natural sciences.

Part I Structure

2

An Overview of the Small-World Phenomenon

2.1 SOCIAL NETWORKS AND THE SMALL WORLD

The "small-world phenomenon" has long been an object of fascination and anecdotal report. The experience of meeting a complete stranger with whom we have apparently little in common and finding unexpectedly that we share a mutual acquaintance is one with which most of us are familiar. More generally, most people have at least heard of the idea that any two people, selected randomly from almost anywhere on the planet, are "connected" via a chain of only a few intermediate acquaintances. Ouisa, a character in John Guare's play *Six Degrees of Separation* (1990), famously claims that

> Everybody on this planet is separated by only six other people. Six degrees of separation. Between us and everybody else on this planet. The president of the United States. A gondolier in Venice . . . . It's not just the big names. It's anyone. A native in a rain forest. A Tierra del Fuegan. An Eskimo. I am bound to everyone on this planet by a trail of six people. It's a profound thought . . . . How every person is a new door, opening up into other worlds.

"Six degrees" is now firmly embedded in folklore, embracing everyone from Kevin Bacon to Monica Lewinsky (Kirby and Sahre 1998). As a result, the idea can be hard to take seriously. And yet, whatever the precise number, it seems that human social systems really are constructed in a fashion quite unlike that of physical systems, in that they seem to violate what is known as *transitivity* of distances. In physical systems (which we generally visualise in no more than a three-dimensional space) all lengths between points, objects, or subsystems are related to each other by the triangle inequality which states that if three points (a, b, and c) are anywhere in the same space, then they can be connected via the three sides of a triangle, and the lengths of those sides must obey the inequality $d(a, c) \leq d(a, b) + d(b, c)$. It seems that this need not be true of social systems, because it is quite possible for person A to be well acquainted with both person B and person C, yet for B and C to be not even remotely familiar with each other. This is a normal part of life, as each of us belongs not to a single group of acquaintances but to many, within each of which everyone pretty much knows everyone else but between which little interaction occurs. Common as it is, this feature of interpersonal relationships has much to do with everyone in the world's being somehow "close" to everyone else, no matter how far away intuition would suggest they are.

2.1.1 A Brief History of the Small World

Theoretical Work

Research specific to the small-world phenomenon did not commence until the 1960s with the formulation and initial mathematical investigation of the problem by Manfred Kochen and Ithiel de Sola Pool (Pool and Kochen 1978). These authors made substantive progress on the problem, estimating both the average number of acquaintances that people possess and the probability of two randomly selected members of a society being connected by a chain of acquaintances consisting of one or two intermediaries. They developed these approximations under a variety of assumptions about the level of social structure and stratification present in the population and concluded (speculatively) that even quite structured populations would have acquaintance chains whose characteristic path lengths are not much longer than those of completely unstructured populations (where the probability of A knowing C, given that A knows B, is independent of whether or not B knows C). For a population about that of the United States and an estimated average number of acquaintances per person of about a thousand, Pool and Kochen estimated that any member of the population could be connected to any other with a chain of associates consisting of at most two intermediaries (hence three degrees of separation).[1]

The study of distances in social networks, however, had begun over twenty-five years before the publication of Pool and Kochen's work, with Anatol Rapoport and his colleagues at the University of Chicago. In a series of papers in the 1950s and 1960s, published in the *Bulletin of Mathematical Biophysics*, Rapoport and colleagues established the theory of *random-biased nets*, which describes the statistics of a disease spreading through populations with varying degrees of structure. Initially, Solomonoff and Rapoport (1951) developed the idea of dispersion in a randomly connected network, in which every element was assumed to have the same number of connections (k). Based on these assumptions—the *independence* of the connections and the *regularity* of the nodes—they derived approximate formulae for the expected fraction (η) of the population to be reached eventually from a small, initial starting set. If a fraction $P(0) \ll 1$ of the population is infected initially, then $(1 - P(0))$ is initially *uninfected*. A consequence of the independence condition is that the disease spreads exponentially, infecting $P(t)$ new members at each time step, where

$$P(t) = \left(1 - \sum_{i=0}^{t-1} P(i)\right)\left(1 - e^{-kP(t-1)}\right). \tag{2.1}$$

In the limit of large t, Rapoport determined the following expression for the total infected fraction η:

$$\eta = 1 - (1 - P(0))e^{-k\eta}. \tag{2.2}$$

Several advances in realism have been made since upon this approximation, most notable of which are

1. The restriction to a *finite subpopulation* from which the k acquaintances can be chosen and a corresponding *strong overlap of friendship circles* (Rapport 1953a, 1953b).
2. The introduction of *structural biases*, specifically, *homophily* (the tendency to associate with people "like" yourself), *symmetry* of edges (which implies undirected instead of directed edges), and *triad closure* (the tendency of one's acquaintances to also be acquainted with each other; Foster et al. 1963).
3. *Social differentiation* of a population into heterogeneous subgroups (Skvoretz 1989).

In later work Rapoport (1957) accounted for the fact that, in a real network, each infected person may have *contact* with k others, but some of these will already have been infected, so that the *effective* number of connections per member t steps away from the origin of the disease is actually $\kappa(t) \leq k$, for all $t > 0$, and is no longer a constant. Fararo and Sunshine (1964) and Skvoretz (1985) argued for a constant $\kappa(t) = \kappa$, but one that accounts for structural biases, yielding the expression (from Skvoretz 1989)

$$P(t) = \left(1 - \sum_{i=0}^{t-1} P(i)\right)\left(1 - e^{-\kappa P(t-1)}\right), \tag{2.3}$$

where, in the special case of undirected ties,

$$\kappa = \begin{cases} \frac{(k-1-\zeta(k-1))(1-(1-\zeta^2)^k)}{k\zeta^2}, & \zeta > 0 \\ k-1, & \zeta = 0 \end{cases}$$

in which

$\zeta = \zeta_s(1 - qS)$ is the "triad-closure bias"
$\zeta_s =$ the triad-closure bias for "strong ties"
$\zeta_w =$ the triad-closure bias for "weak ties"
$S = (1 - (\zeta_w/\zeta_s))$ is a measure of the "strength of weak ties"
$q =$ the probability of a connection being weak.

Note that the above expression requires a distinction between "strong" and "weak" ties, where the *strength* of a tie is determined not by some inherent feature of the tie itself, but by the structure of the surrounding network. Specifically, Granovetter (who introduced the idea) defined strength as follows:

> Consider, now, any two arbitrarily selected individuals—call them A and B—and the set, $S = C, D, E, \ldots,$ of all persons with ties to either *or* both of them. The hypothesis which enables us to relate dyadic ties to larger structures is: the stronger the tie between A and B, the larger the proportion of individuals in S to whom they will *both* be tied, that is, connected by a weak or strong tie. This overlap in their friendship circles is predicted to be least when their tie is absent, most when it is strong and intermediate when it is weak. (1973, p. 1362)

In a later article (1983), Granovetter stresses that weak ties are, in fact, more significant in a social network than their strong counterparts:

> The argument asserts that our acquaintances ("weak ties") are less likely to be socially involved with one another than are our close friends ("strong ties")....
>
> The overall social structural picture suggested by this argument can be seen by considering the situation of some arbitrarily selected individual—call him or her "Ego." Ego will have a collection of close-knit friends, most of whom are in touch with one another—a densely knit "clump" of social structure. In addition, Ego will have a collection of acquaintances, few of whom know one another. Each of these acquaintances, however, is likely to have close friends in his or her own right and therefore to be enmeshed in a closely knit

> clump of social structure, but one different from Ego's. The weak tie between Ego and his or her acquaintance, therefore, becomes not merely a trivial acquaintance tie, but rather a crucial bridge between the two densely knit clumps of close friends. To the extent that the assertion of the previous paragraph is correct, these clumps would not, in fact, be connected to one another at all were it not for the existence of weak ties. (p. 203)

The probability (q) that a tie is weak corresponds to the likelihood that two connected vertices will have weakly overlapping friendship circles, and S quantifies how much less likely a weak tie is to complete a triad than a strong tie. Hence S is a measure of how significant a role the weak ties will play. The importance of the "strength of weak ties" idea was reinforced by Skvoretz and Fararo (Skvoretz 1989), who showed that "the stronger the weak ties in a population (in two senses, weak tie triads being less likely to close and weak ties being proportionately more frequent), the closer is a randomly chosen starter to all others."

Exactly how this works, and how important it is, is a major concern in this book. Closely related to the strength of ties and triadic closure is the idea of clustering, which has also been an issue of concern to researchers in social networks and has significant connections to the small-world problem. The idea of networks being divisible into cooperative subgroups that do not cooperate with each other was first formulated by Davis (1967), but the idea that neighbourhoods could be more or less densely connected was not quantified until slightly later by Barnes (1969), who defined *density* at a network element v as the proportion of all possible connections in v's immediate neighbourhood (defined by v and those elements to which v is connected directly) that actually exist. A very similar notion of density, termed *clustering*, is defined later in this chapter[2]. Barnes also discussed some qualitatively different networks for the parameters $n = 100$, $k = 10$, observing that they have different local densities and that the typical separation of network elements appears to increase with increasing density. Whilst this analysis touched on the idea that the local properties of a network (like density) can determine its global properties, the first systematic attempt to relate the two scales appears to have been the development of the concept of *structural eqivalance* and the technique of *block modelling*. According to Lorrain and White (1971), "a is structurally equivalent to b if a relates to every object x of [a category] C in exactly the same way as b does. From the point of view of the logic of the structure, then a and b are absolutely equivalent, they are substitutable."

Hence clustering and structural equivalence capture much of the same information about "who knows whom" at a local level, at least in the restricted case in which only one kind of social relation is considered. Block modelling (White et al. 1976) then considers networks as composed of *blocks* of structurally equivalent elements and represents the graph in terms of the relationships between these blocks. This is analogous to the amalgamation of Barnes's clusters connected by Granovetter's weak ties as a means of constructing a global view of the network whilst retaining some knowledge of the local structure. It is different from the work of the biased-net theorists because, instead of considering the characteristics of *pathways* though networks, it looks directly at the *knittedness* of networks at different scales.

A final area of research in social networks that relates closely to the work presented here is that concerning the dimension and geometry of the space in which social networks are presumed to exist. Many approaches to this question have been devised, but almost all of them fall under the rubric of *multidimensional scaling*. This term refers to a loose bundle of techniques developed by many researchers across several disciplines and decades, but all are based on more or less the same idea, which is basically the following.

A population is assumed to exist in some finite (but possibly large) dimensional "social space," where the coordinates $(x_1, x_2, \ldots, x_m)$ of each member represent quantitative measures of a set of characteristics, which are presumably sufficient to identify each member of the population uniquely. These coordinates, however, are unknown to the observer, as is the dimension of the space. What the observer *does* know is the *distances* $(\delta_{i,j})$ between each pair (i, j) of members, where distance is defined in some manner that is problem-specific but is often related to frequency of interaction or some assessment of similarity, generated by either the observer or the members themselves. The problem then is to reconstruct the space by choosing coordinates in a self-consistent manner such that the known distances are related to the coordinates by a particular choice of metric. Basically, it is at this point that the methods start to differ.

The group of methods generally known as *metric methods* (Chapter 4 of Davidson 1983) utilise a standard Euclidean metric, hence

$$\delta_{i,j} = \left[\sum_d (x_{id} - x_{jd})^2 \right]^{\frac{1}{2}}, \tag{2.4}$$

whereas an alternate group of methods known as *nonmetric methods* (Chapter 5 of Davidson 1983) utilise a variant of Equation 2.4:

$$\delta_{i,j} = f\left(\left[\sum_d (x_{id} - x_{jd})^2\right]^{\frac{1}{2}}\right), \tag{2.5}$$

where f is some monotone function (that is, $d_{ij} < d_{i'j'} \Rightarrow f(d_{ij}) < f(d_{i'j'})$). In fact, "nonmetric" is something of a misnomer, because Equation 2.5 is every bit as much a metric as Equation 2.4; it is just not the Euclidean metric. It turns out that this confusion between nonmetric and non-Euclidean is widespread in this particular part of the study of social networks. For instance, Barnett (1989) claims that non-Euclidean geometry is necessary to describe social networks precisely. Pool and Kochen also believed this, asserting that the transitivity of a Euclidean space is violated in social networks where Person A may be very close to both B and C and therefore likely to know them both, but B and C may be very far from each other (Pool and Kochen 1978). The conclusion is that if this basic tenet of a Euclidean space is violated, then, necessarily, the space in which social networks exist must be non-Euclidean. This is actually a misunderstanding about which more will be said in Section 2.1.2.

For the moment, it is important to realise only that multidimensional scaling is simply a process designed to reconstruct the space in which the system is presumed to exist, thus both generating a set of meaningful coordinates, with which to distinguish population members, and providing a visual representation of the data that enables the observer to gain more insight into the relationships between members than would be possible by staring at a large matrix of numbers. Therefore, in applying to the data whichever particular algorithm is chosen, it is to be hoped that only a few dimensions will be sufficient to embed the data within an acceptable degree of error. Obviously, if $\Delta_{i,j}$ is an $n \times n$ matrix containing the intermember distance information, it is always possible to embed the network in an n-dimensional space. However, for $n > 3$, this isn't going to be much help in visualising the data relationships expressed in the matrix, and from another perspective, there is almost no point in doing it at all (for any n) because the resulting relationships between the coordinates will be no less impenetrable than the original distance matrix. Hence there is a substantial trade-off between the goodness of fit of the embedding and its dimension, which should almost always be kept to less than four. This then raises another thicket of issues for the data analyst, but these are not relevant here. See Davidson (1983) and R. N.

Shepard and Nerlove (1972) for an overview of these obstacles and the methods attempted to surmount them.

The theory of social networks, then, has proceeded along four basically distinct, but interrelated, strands:

1. The statistical analysis of pathways through networks with varying degrees of local structure.
2. The qualitative description of the structure of networks in terms of local (e.g., clustering) and nonlocal (e.g., weak ties) features.
3. The renormalisation of networks, viewed as meta-networks of highly clustered or equivalent subnetworks.
4. The embedding of networks into (hopefully low-dimensional) spaces where the coordinates are readily interpretable and relationships between members can be more easily visualised.

In tandem with (and frequently driving) this theoretical development has been the development of empirical techniques that attempt to probe the structure of real social networks directly. Once again, the small world did not start turning in this field of endeavour until the late 1960s.

Empirical Work

The first empirical work was conducted at about the same time as Pool and Kochen were developing their theoretical ideas, by the psychologist Stanley Milgram. Although principally renowned for his remarkable and disturbing work on the apparent submission of human ethical values to authority (Milgram 1969), Milgram also conducted a highly innovative test of the small-world hypothesis (Milgram 1967). In this experiment, Milgram sent a number of packets to agreeable "sources" in Nebraska and Kansas, with instructions to deliver these packets to one of two specific "target" persons in Massachusetts. The targets were named and described in terms of approximate location, profession, and demography, but the sources were only allowed to send the packets directly to someone they knew by first name. The object was to get the packets from source to target with as few of these "first-name-basis links" as possible. Hence each link in the chain was required to think hard about which of their acquaintances would be most likely to know the target person or at least be "closer" to them: demographically, geographically, personally, or professionally. Also, each link was supposed to record, in the packet, details about themselves corresponding to those provided about the target, enabling the experimenters to track the progress of the packet and the demographic nature of chain along which it passed.

The upshot of all this was that Milgram determined that a median of about five intermediaries was all that was required to get such a letter across the intervening expanse of geography and society. Whether this number is, in reality, too low or too high is a matter of debate. On the one hand it would seem unlikely that, at every step, the sender would pick the optimal person to send it to (and they could only pick one), and that this effect would tend to make the chains longer than they needed to be. On the other hand, many of the chains were never completed because of apathy on the part of the participants, and, as longer chains are more likely to terminate than shorter ones, the result might well have been systematically biased in favour of lower numbers. White (1970) proposed a model to account for this effect, which yielded a revised estimate of about seven intermediaries. In any event, Milgram seemed to have demonstrated that whatever the precise number was, it wasn't very big, compared with the overall magnitude of the system (on the order of the population of the United States, which was about 200 million in 1967). A second study by Milgram (Korte and Milgram 1970) used essentially the same method to examine the length of acquaintance chains between whites in Los Angeles and a mixed white-black target population in New York and found similar statistics.

Of course, the study of social networks and their use as a tool for examining the structure of societies already had a considerable history by the time Milgram did his initial experiment (see Mitchell 1969 for a review of the field at the end of the 1960s), but none of this work had looked at the question of *path length* in the same light as had Milgram. It also seems that very little work of this nature and scale has occurred since, even though Milgram's results did (and still do) spur considerable interest. Perhaps the work of Rapoport (Foster et al. 1963) is closest to this, in that he measures the average fraction of a population of students in a junior high school that is reached as a function of number of intermediaries. Even here, though, the system involved is much smaller, and the emphasis is upon justifying parameters for a network model rather than direct experimental verification of the small-world phenomenon.

In fact, it seems that more empirical effort has been devoted to the lower-level question (originally posed by Pool and Kochen 1978) of the number of acquaintances that the typical person possesses. Efforts in this department have been made by Freeman and Thompson (1989), who use a variant of Pool's original "telephone book" method,[3] and Bernard et al. (1989), who use the 1985 Mexico City earthquake victims as a sample subpopulation to determine the acquaintance volume of residents of Mexico City. This turns out to be a difficult exercise, and it seems unlikely that even if such a number and its variance could be convincingly determined for any given definition of acquaintance, that it would play nearly so as important a role in the understanding of networks as a comparable advance in the understanding of network structure.

2.1.2 Difficulties with the Real World

Theoretical

Although the researchers surveyed in Section 2.1.1 did make significant gains on the issue of the effective size of social networks, their progress was hampered by a number of difficulties that arose from both the questions they chose to ask and the methodologies they used to seek answers. The results of Pool and Kochen are highly suggestive of the small-world property's holding true in real societies. But although their results are not highly sensitive to the estimation of the average number of acquaintances,[4] they *are* highly sensitive both to the assumptions about conditional probability of acquaintanceship and to the large-scale structure of the population, which may dictate different rules of conditional probability in different parts of the population. A more recent article by Kochen (1989b) reports little progress on this essential theoretical difficulty. It turns out that this is a problem faced by all theoreticians who find themselves exploring systems that operate in the intermediate regime between order and randomness. The problem arises in many fields, notably fluid mechanics and the dynamics of coupled, nonlinear oscillators (see Chapter 9), but in terms of social networks, the only networks whose statistical properties are analytically tractable are those that are either (1) completely ordered (for instance, a d-dimensional, hypercubic lattice), or (2) completely random (such as Rapoport's random webs).

Although these cases are at opposite extremes of the structural spectrum, they both share the essential characteristic that their local structure mirrors (either exactly or statistically) their global structure, and hence analysis based on strictly local knowledge is sufficient to capture the statistics of the entire network. That is, in an important sense, they "look" the same everywhere.

Unfortunately, real social systems appear to be firmly in between these extremes, and, to make matters worse, it is not even known where on the spectrum they lie. What does seem to be true is that, if any theoretical explanation is to capture the important features of social networks, then it must find some way of encapsulating elements of both order and randomness, thus accounting for the appearance of structure at different scales. Much of the work surveyed above has grappled with this problem in creative and insightful ways, but three central issues appear to remain open:

1. Social networks exhibit structural characteristics that are inherently *nonlocal* (Granovetter's "bridges"), and so no purely local analysis can predict their global statistical features.
2. Analytical difficulties increase with the size of the network, and almost none of the work has been tested for large population size (n) with sparse connectivity under any but the most restrictive conditions.
3. It is unknown where on the structural spectrum real social networks lie, but no treatment has been given to the properties of *continuous families of networks*, whose structural properties vary all the way from one extreme to the other, with the intention of determining the location and nature of any *transitions* that occur in between.

Adding to the confusion is the difficulty of determining which kind of space a network exists in and the appropriate metric with which to measure lengths. The root of this difficulty appears to be that networks are frequently defined in the sociological literature on the basis of (at least) two relations: (1) how "far" each pair of vertices is from each other in the (unknown) metric of the (unknown) "social space," and (2) whether or not they are connected and (perhaps) how strongly.

The first relation turns out to be the problematic one because if one takes any single measure of "social distance," such as frequency of interaction, overlap of interests, or common characteristics, ambiguities inevitably arise, and the resulting "distances" will appear to violate the triangle inequality. It is false, however, to declare the corresponding space non-Euclidean. In fact, the violation of the triangle inequality (if it isn't just due to faulty data) is symptomatic of a far more general breakdown in the geometry of a space, because it violates one of the fundamental notions not just of Euclidean distance, but of distance itself. The reason for this is that the triangle inequality is one of four basic properties of a class of spaces known as *metric spaces* (Munkres 1975). This is an extremely general class of topological space that formalizes the idea of distance (that is, a *metric*) and that includes *any* sensible notion of distance. Hence if the measured "distances" in some network are not consistent with the triangle inequality, then either (1) the criteria used to measure distance are mistaken (the data are somehow incomplete or in error), or (2) the space is not a metric space, and so the concept of distance is meaningless in the first place.

In either case the result is that not much can be done to interpret the measured distances as meaningful without performing some arbitrary (and probably meaningless) transformation on them (such as adding a large constant to all distances or taking logarithms) until they *do* satisfy the requirements of a legitimate metric. Such manipulations, however, do not so much help the data as reveal their inherent flaws and suggest either a new method of measuring social distance or a different approach to the problem altogether.

The most likely source of the problem is the difficulty inherent in measuring how close or far apart people are, not in a network, but in the more general sense implied by the idea of a social space. As soon as one tries to grapple with this issue, it immediately reveals itself to be slippery both theoretically (what does "social distance" mean and what are its most important contributors?) and empirically (even if you knew what it was, how would you go about measuring it?). Even in the best-case scenario, it seems that whatever metric of length is chosen, it will almost certainly not capture all the features that are relevant to relationships between people. It is just as likely, however, that such a measure could even be *multivalued*, as social distance is at least partly a matter of perspective.

In the case of a network, the issue becomes still murkier, because distance can also be defined in terms of the network connections themselves, which may be a function of the underlying space but almost certainly not one that is known. If the network distance and the metric distance do not agree, then the analyst is once again faced with a choice between arbitrary manipulations of the data and ignoring them outright. Because the methodological basis of measuring distances in the network sense, solely in terms of who is connected to whom, rests on much firmer ground, both theoretically and empirically, network distance will be treated here as the sole measure of distance, at which point all talk of either non-Euclidean or nonmetric spaces instantly disappears. A network does not necessarily exist in any particular space at all, but as all network distances must certainly conform to the triangle inequality, then (if one insists on thinking in terms of Euclidean spaces) an embedding is guaranteed by an algorithm that is described briefly in Section 2.2.3.

Empirical

On the empirical side of the same problem, the principal stumbling block seems to have been the practical difficulties associated with obtaining and representing sufficiently detailed relationship data for large, sparse networks. Milgram's methodology (sometimes referred to as "the small-world method") was imaginative and original and did serve to illuminate some interesting characteristics of what might be termed a "random-biased walk" in a network: random, because senders did not have sufficient information to know which of their acquaintances was optimal, and biased, because some attempt at optimality was made. Unfortunately it is difficult to generalise results such as these beyond the scope of the specific study, and so it is hard to tell much about the overall qualitative structure of the network.

Conversely, attempts to reconstruct network connectivity in a broader sense have been forced to concentrate on small systems in which it is practicable to map every single connection (see, for example, Doreian 1974), at which point there is little of the intermediate ground between the local and the global scales in which the interesting small-world phenomena occur.[5]

Finally, the methods used to assess actual network parameters, such as the average number of friends per person, have revealed that deep problems exist with any attempt to estimate this kind of data:

1. Most people seem to be quite poor at estimating their number of friends reliably.
2. Methodological tricks to circumvent this difficulty (such as requiring subjects to keep written records of all interpersonal encounters over an extended period of time) are time and labour intensive.
3. The number (however it is estimated) changes over time.
4. The number is highly sensitive to the definition of a "meaningful" or "substantive" contact or relationship.

Of all these points, perhaps the last is the most damning because it threatens the validity of any result that does not perform at least several identical surveys of acquaintanceship volume using different definitions of what constitutes acquaintanceship in each (for instance, first-name basis versus propensity to lend money). This is a similar objection to that raised earlier concerning the definition of "social distance": acquaintanceship, like distance, can vary widely depending on any one or all of the following: (1) the biases of the observer, (2) the question being posed, and (3) the members of the network in question.

2.1.3 Reframing the Question to Consider *All* Worlds

Given the difficulties inherent to empirical investigations of the small-world phenomenon (and the structure of social networks in general), theoretical investigation seems attractive, if only as a means of focusing future questions for empiricists to answer. Theoretical approaches, however, also seem to have some serious limitations in the regime of interest, at least if one insists upon analytical solutions. What is needed, then, is a new theoretical approach that attempts to exploit the generality of theory without falling prey either to overly restrictive idealisation of network structure ("throwing the baby out with the bathwater") or to impenetrable thickets of numerical solutions in a forest of arbitrary parameters. The motivation behind the work presented here is to chart just such a course by ignoring much of the sociological detail inherent in previous models and considering a much more general statement of the problem:

> Assuming that a network can be represented by nothing more than the connections existing between its members and treating all such connections as equal and symmetric, a broad class of networks can be defined, ranging from highly ordered to highly random. The question then is *Does the Small-World Phenomenon arise at some point in the transition from order to disorder, and if so, what is responsible for it?* In other words, *What is the most general set of characteristics that guarantee that a system will exhibit the small-world phenomenon, and can those characteristics be specified in a manner independent of the model used to create the network?*

As yet, the small-world phenomenon has not been defined precisely in terms of which specific properties a network must possess in order to exhibit it. This will be deferred until after some exploration of different network topologies yields the kind of intuition that will be needed to motivate the appropriate definition. Even so, it should be apparent that if the goal stated above can be achieved, a great deal can be said about the existence of the small-world phenomenon, in what systems it is likely to arise, and in what sort of applications it might be useful. Before continuing, however, some basic terminology and results are required from the theory of graphs that will help to describe and define the networks and properties of interest.

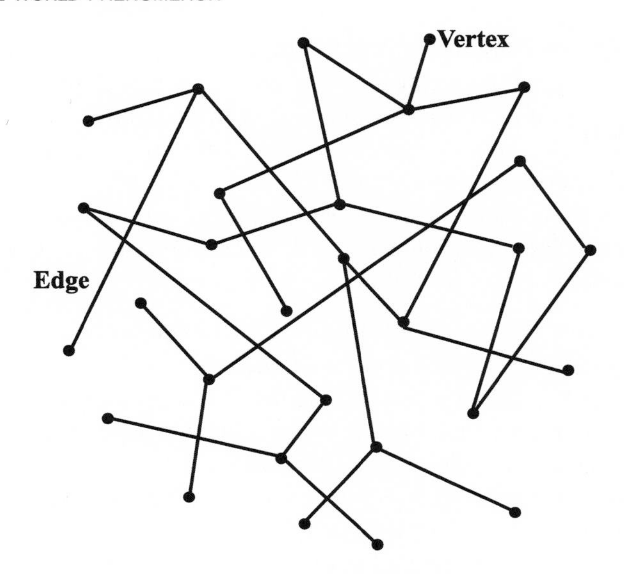

Figure 2.1 A general graph.

2.2 BACKGROUND ON THE THEORY OF GRAPHS

2.2.1 Basic Definitions

A graph, in its most basic sense, is nothing more than a set of points connected in some fashion by a set of lines (see Fig. 2.1). The following definition of a graph is taken from Wilson and Watkins (1990).

Definition 2.2.1. A *graph* G consists of a nonempty set of elements, called *vertices,* and a list of unordered pairs of these elements, called *edges*. The set of vertices of the graph G is called the *vertex set* of G, denoted by $V(G)$, and the list of edges is called the *edge list* of G, denoted by $E(G)$. If v and w are vertices of G, then an edge of the form vw is said to *join* or *connect* v and w.

The number of vertices in $V(G)$ is termed the *order* of the graph (n), and the number of edges in $E(G)$ is termed its *size* (M). Graphs can be used to represent all kinds of networks, where the vertices represent network elements (such as people, animals, computer terminals, organisations, cities, countries, production facilities) and the edges represent some predefined relationship between connected elements (such as friendship, prey-predator relationship, Ethernet connection, business alliance, highway, diplomatic relationship, product flow). Clearly both elements and connections may embody any number of characteristics, but the theory of graphs generally deals only with the number of elements in the network and their relationships with respect to each other in terms of the characteristics of the edge set.

This very broad definition is capable of representing systems in bewildering detail. The only graphs that will be considered here, however, are those that conform to the following restrictions:

1. *Undirected*. Edges exhibit no inherent direction, implying that any relationship so represented is symmetric.
2. *Unweighted*. Edges are not assigned any a priori strengths. Hence any importance that specific edges may later assume derives solely from their relationship with other edges.
3. *Simple*. Multiple edges between the same pair of vertices or edges connecting a vertex to itself are forbidden.
4. *Sparse*. For an undirected graph, the maximal size (M) of $E(G) = \binom{n}{2} = n(n-1)/2$, corresponding to a "fully connected" or *complete* graph. Sparseness implies $M \ll n(n-1)/2$.
5. *Connected*. Any vertex can be reached from any other vertex by traversing a path consisting of only a finite number of edges.

These assumptions obviously compromise the ability of the resulting models to form realistic representations of many networks. Relationships are often directed (child-parent, teacher-student, and so on), and some are clearly more important than others; many real networks are not connected, and often multiple types of relationships exist between the same set of elements (like business and friendship ties). However, aside from simplifying the resulting analysis a great deal, these assumptions do form a natural starting point for modelling networks in that they introduce a minimum amount of arbitrary structure whilst still allowing meaningful questions to be asked of the network as a whole.

Although graphs can be represented pictorially, most computations of graph properties are accomplished by way of either an *adjacency matrix* or *adjacency list*. The *adjacency matrix* $\mathbf{M}(G)$ is the $n \times n$ matrix in which $M_{i,j}$ is the number of edges joining the vertices i and j. In the unweighted case, all entries would be either 0 or 1. The *adjacency list* simply lists all vertices of the graph and, next to each, the vertices with which it is adjacent. The number of edges incident with a given vertex v (that is, the size of v's adjacency list) is called the *degree* of v, denoted k_v. One statistic that will be referred to frequently is the *average degree* of the graph, k. Hence, for undirected graphs, k quantifies the relationship between n and M ($M = (n \cdot k)/2$). The corresponding effect on k of the sparseness condition above is that all graphs must have $k \ll n$. A graph in which all vertices have precisely the same degree k is called *k-regular* or just *regular*.

2.2.2 Length and Length Scaling

One of the most important statistics of graphs to be considered here is the *characteristic path length* ($L(G)$), that is, the typical distance $d(i, j)$ between every vertex and every other vertex. "Distance" here refers not to any separately defined metric space in which the graph has been embedded, but to a distinct graph metric—simply the minimum number of edges (in the edge set) that must be traversed in order to reach vertex j from vertex i, or in other words the *shortest path length* between i and j. Investigations of this graph invariant have a long history, spanning several subject areas and utilising a number of approaches. As long ago as 1947, Wiener (1947) investigated the sum of all distances between all pairs of vertices in a graph (sometimes called the "Wiener index") in connection with the boiling point of paraffin, where the vertices of the graph were to represent atoms and the edges, intramolecular bonds. Since then both the sum of all distances in a graph as well as the average distance across all pairs of vertices have appeared as parameters relevant to social status in a hierarchy (Harary 1959), architectural floor plans (March and Steadman 1971), the performance of computer networks (Frank and Chou 1972) and telecommunication networks (Lin 1982; Pippenger 1982; Chung 1986), and the physical properties of as yet unsynthesised hydrocarbons (Rouvray 1986).

Throughout all this the problem of finding a closed-form expression for the characteristic path length of a general, connected graph[6] has remained impregnable, and researchers have had to satisfy themselves with the explication of upper and lower bounds upon the quantity for various classes of graphs. Cerf et al. (1974) determined a lower bound on the average $d(i, j)$ in a k-regular graph by assuming a perfectly expanding graph. That is, starting from any vertex, k vertices can be reached at distance 1, then from each of these vertices another $(k - 1)$ new vertices can be reached at distance 2, and so on, without any redundancies, until the entire graph has been reached. This type of graph, known as a Moore Graph, is the most *efficient* possible k-regular graph, in the sense that every vertex "reaches" $(k - 1)$ new vertices, but it has since been proven unrealisable except in a handful of special cases in which it is possible to close the graph upon itself with no redundancies (Chung 1986). An important result of studying Moore Graphs as a theoretical lower bound (even if unattainable) is that, for $k > 2$, the characteristic path length in any *regular* graph must grow at least logarithmically with n. We will see later that random graphs are good approximations to this lower bound.

Entringer et al. (1976) showed that the sum of all distances in any graph must lie between that of a complete graph and that of a one-dimensional *chain,* where each internal vertex has $k = 2$. Doyle and Graver (1977) later showed that a cycle—a chain with its ends connected—has the maximal characteristic path length of any graph with periodic boundaries. Whilst this result does not necessarily extend to higher k (where vertices are connected to nearest neighbours, next-nearest neighbours, and so on), it suggests that cycles with larger k have close to the maximum possible characteristic path length for a given n and k. It also suggests that the cycle is a particularly interesting object because it is, at once, the most and least efficient 2-regular structure. In fact, it is the *only* 2-regular structure, and has the additional property that it is also 2-connected, which is to say that the deletion of any two edges will disconnect the graph. Hence it is the only minimally connected, regular graph topology—a fact that will be useful in Chapter 3.

Following this work, tighter bounds on the average distance or sum of all distances have been determined for specific classes of graphs (Buckley and Superville 1981), digraphs (Plesnik 1984), trees (Winkler 1990; Entringer et al. 1994), and random graphs (Schneck et al. 1997). The greatest problem with these attempts is that they either impose very loose bounds on the quantity of interest or else require strong constraints on the class of graphs to which the bounds apply. In either case, the results do little to assist the task of actually determining a characteristic path length for an arbitrary graph.

Recently, some new approaches have been developed by (amongst others) Chung (1988, 1989, 1994) and Mohar (1991), which place bounds on the characteristic path length without restricting the variety of eligible graphs. Unfortunately these bounds are necessarily expressed in terms of other graph invariants that are virtually as inaccessible as the characteristic path length itself.[7] Interesting though such relationships between graph invariants are, they do not really help much if the primary aim is specifically that of computing or estimating length. They also suggest that analytical formulae for the length characteristics of graphs are, in general, hard to come by. Hence a heavy reliance on numerical results seems appropriate.

At this point, it might seem that the obvious and natural definition of characteristic path length would be $d(i, j)$ averaged over all $\binom{n}{2}$ pairs of vertices and that this is best computed numerically for a known graph. Unfortunately, for large n, this becomes impractical to compute exactly, so a random sampling technique is needed to estimate the length to within a prescribed accuracy. Using such a sampling technique, it turns out that it is significantly easier to estimate the *median* shortest path length than it is the mean. As the mean and the median are practically identical for any reasonably symmetric distribution, then the sampling efficiency of the median seems to mark it as the most appropriate measure of length in a graph. However, the median suffers from a different drawback, which is that it is integer-valued. As the *scaling properties* of length with respect to increasing n are also of interest, and as the characteristic path length of some graphs remain on the same order of magnitude over several orders of magnitude in n, then an integer-valued length cannot provide sufficiently detailed information. A reasonable compromise, which incorporates most of the sampling convenience of the median, with the real-valued advantage of the mean is the following.

Definition 2.2.2. The *characteristic path length* (L) of a graph is the *median* of the *means* of the *shortest path lengths* connecting each vertex $v \in V(G)$ to all other vertices. That is, calculate $d(v, j)\ \forall j \in V(G)$ and find $\bar{d}_v$ for each v. Then define L as the median of $\{\bar{d}_v\}$.

As mentioned above, for large n a random sampling technique is used that is due to Huber (1996). According to this method, $\bar{d}_v$ is calculated for a randomly selected subset of s vertices, where s is determined as follows:

> Finding an approximate median through sampling is relatively straightforward. First, take s samples, then find the median of the samples. More generally, call $M_{(q)}$ a *q-median* if at least qn of the numbers in the set (n) are less than or equal to $M_{(q)}$ and at least $(1-q)n$ of the numbers are greater than or equal to $M_{(q)}$.
>
> Call $L_{(q,\delta)}$ a *(q, δ)-median* if at least $qn(1-\delta)$ numbers in the set are less than or equal to $L_{(q,\delta)}$ and at least $(1-q)n(1-\delta)$ of the numbers are greater than $L_{(q,\delta)}$. Equivalently, $L_{(q,\delta)} = M_{(q')}$ for some q' that satisfies $(1-\delta)q \leq q' \leq (1+\delta)q$.
>
> Finding such a value $L_{(q,\delta)}$ that is correct with high probability is much faster than finding M_q which takes linear time. To find a value

for $L_{(q,\delta)}$, take s samples and look at the q-median of the sample (p. 2).

Theorem 2.2.1. *The above algorithm yields a correct value for $L_{(q,\delta)}$ with probability $1-\varepsilon$ if s samples are taken, where $s=(2/q^2)\ln(2/\varepsilon)1/(\delta')^2$ and $\delta'=1/(1-\delta)-1=\delta/(1-\delta)$. Note that when δ is small, $\delta\simeq\delta'$* (Huber 1996).

The computation required by Definition 2.2.2 is less efficient than that actually proposed by Huber, which samples only s pairs of vertices instead of s complete search trees. However, the difference in computational time is only a constant factor and so is a reasonable sacrifice to make for the utility of a real-valued measure of length.

Having established either an exact (convenient in practice only for $n \lesssim 1{,}000$) or approximate value of the characteristic path length, the question arises: how does L *scale* with respect to changes in n and k? This question is important because the *scaling* of L is more indicative of the *qualitative structure* (or *topology*) of a graph than the specific value of L itself. Precisely what is meant by maintaining the qualitative structure of a graph whilst changing n and k will become more apparent in Chapter 3 in terms of one-parameter families of graphs (parameterised by some parameter p) that interpolate between order and randomness. The point is that different values of the parameter p represent different qualitative structures and that graphs of different n and k, but with the same p value, are qualitatively the same. This leads to the following definitions.

Definition 2.2.3. For a fixed p, the *length scaling with respect to n of* $G(p)$ is

$$\lim_{\substack{n_1\to\infty\\ n_2\to\infty}}\left(\frac{L(G(p;n_1,k))}{L(G(p;n_2,k))}\right)$$

for $n_1>n_2$ and $1\ll k\ll n_1, n_2$. T is said to exhibit *d-scaling with respect to n* if

$$\lim_{\substack{n_1\to\infty\\ n_2\to\infty}}\left(\frac{L(G(p;n_1,k))}{L(G(p;n_2,k))}\right)=\frac{n_1^{\frac{1}{d}}}{n_2^{\frac{1}{d}}}.$$

Definition 2.2.4. For a fixed p, the *length scaling with respect to k of* $G(p)$ is

$$\lim_{n\to\infty}\left(\frac{L(G(p;n,k_1))}{L(G(p;n,k_2))}\right)$$

for $k_1>k_2$ and $1\ll k_1, k_2\ll n$.

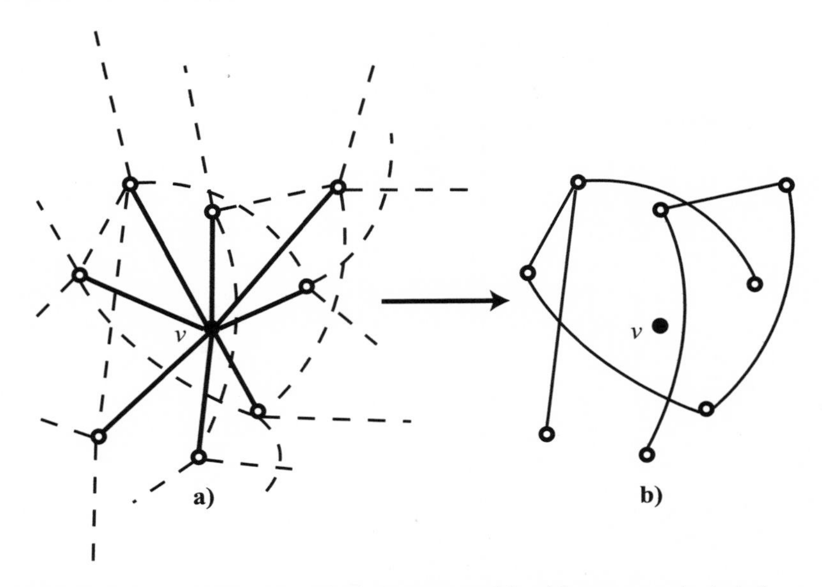

Figure 2.2 The neighbourhood of a vertex v (a) with vertex v included and (b) showing only the edges between vertices in $\Gamma(v)$.

Hence the parameter p defines an infinite set of graphs, each of which exhibits certain structural characteristics common to the set. The characteristic path length of a member of such a set can vary over $1 \leq L < \infty$, but the scaling property is invariant across the entire set. This is why the idea of scaling is so useful in characterising the qualities of a given graph: one can compute exactly the properties of small graphs with a given p and then, knowing their scaling properties, also obtain knowledge of their much larger cousins whose properties cannot be computed directly.

2.2.3 Neighbourhoods and Distribution Sequences

One recurrent theme throughout this book is the metaphor of information "spreading" from a single vertex throughout the graph. In connected graphs, there is no issue of whether or not the entire graph will be reached, but only the number of "steps" required to achieve this. The notion of a step is captured in terms of the *neighbourhood* either of a vertex (see Fig. 2.2) or a connected subgraph.

Definition 2.2.5. The *neighbourhood* $\Gamma(v)$ of a vertex v is the subgraph that consists of the vertices adjacent to v (not including v itself).

TABLE 2.1
Distribution Sequence for Kevin Bacon in the "Kevin Bacon Graph" (April 1997)

j (Bacon number)	$\|\Gamma_j\|$	Λ_j
0	1	1
1	1,181	1,182
2	71,397	72,579
3	124,975	197,554
4	25,665	223,219
5	1,787	225,006
6	196	225,202
7	22	225,224
8	2	225,226

Definition 2.2.6. The neighbourhood $\Gamma(S)$ of a connected subgraph S is the subgraph that consists of all vertices adjacent to any of the vertices in S, but not including the vertices of S.

Definition 2.2.7. In the special case where $S = \Gamma(v)$, $\Gamma(S) = \Gamma(\Gamma(v)) = \Gamma^2(v)$. More generally $\Gamma(\Gamma^{i-1}(v)) = \Gamma^i(v)$, the ith *neighbourhood of* v. Hence $\Gamma^0(v) = \{v\}$.

Definition 2.2.8. The sequence $\Lambda_j(v) = \sum_{i=0}^{j} |\Gamma^i(v)|$ for $0 \leq j \leq j_{\max}$ is the *distribution sequence for* v, where $\Lambda_{j_{\max}}(v) = |G|$.

Definition 2.2.9. $\Lambda_j = \overline{\Lambda_j(v)}$ over all $v \in V(G)$ is the *distribution sequence* for G.

It follows immediately from these definitions that $\max_v(j_{\max}(v)) = D$, the *diameter* of the graph. The functional form of Λ_j is indicative of the rate at which information "spreads" throughout a graph (think of a signal spreading from vertex to vertex along the edges, where all edges take equal "time") and hence the structure of the graph itself. Table 2.1 gives a real example of a distribution sequence for none other than the illustrious Kevin Bacon, where j is the Bacon Number, and Λ_j is the number of actors and actresses who have a Bacon Number of j or less.

2.2.4 Clustering

The idea of a neighbourhood is also useful in quantifying another statistic that will be of interest in this work, namely, the *clustering coefficient* of a graph.

Definition 2.2.10. The clustering coefficient γ_v of Γ_v characterises the extent to which vertices adjacent to any vertex v are adjacent to each other. More precisely,

$$\gamma_v = \frac{|E(\Gamma_v)|}{\binom{k_v}{2}},$$

where $|E(\Gamma_v)|$ is the number of edges in the neighbourhood of v and $\binom{k_v}{2}$ is the total number of *possible* edges in Γ_v.

That is, given k_v vertices in the subgraph Γ_v, at most $\binom{k_v}{2}$ edges can be constructed in that subgraph. Hence γ_v is simply the net fraction of those possible edges that actually occur in the real Γ_v. In terms of a social-network analogy, γ_v is the degree to which a person's acquaintances are acquainted with each other and so measures the *cliquishness* of v's friendship network. Equivalently, γ_v is the probability that two vertices in $\Gamma(v)$ will be connected. Hence a measure of clustering over the entire graph is the following.

Definition 2.2.11. The *clustering coefficient* of G is $\gamma = \gamma_v$ averaged over all $v \in V(G)$. Hence $\gamma = 1$ would imply that the corresponding graph consisted of $n/(k+1)$ disconnected, but individually complete, subgraphs (cliques), and $\gamma = 0$ would imply that no neighbour of *any* vertex v is adjacent with any other neighbour of v.

2.2.5 "Lattice Graphs" and Random Graphs

There are many other graph statistics that could (and probably should) be measured. But already it is possible to make a crude examination of graph structure, starting with some special classes of graphs that will be useful points of reference in Chapters 3 and 4, namely *lattice-graphs* (or *d-lattices*) and *random graphs*.

Properties of d-Lattices

Definition 2.2.12. A *d-lattice* is a labelled, unweighted, undirected, simple graph that is similar to a Euclidean cubic lattice of dimension d in that any vertex v is joined to its lattice neighbours, u_i and w_i, as specified by

$$u_i = \left[\left(v - i^{d'}\right) + n\right] \pmod{n},$$

$$w_i = \left(v + i^{d'}\right) \pmod{n},$$

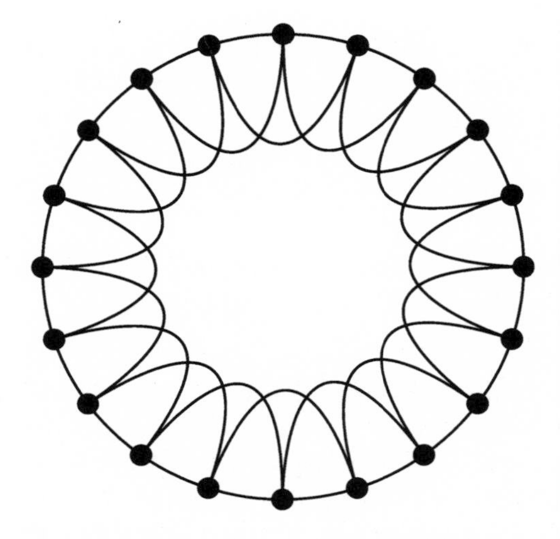

Figure 2.3 Example of a 1-lattice with $k = 4$.

where $1 \leq i \leq k/2$, $1 \leq d' \leq d$, and it is generally assumed that $k \geq 2d$.

Hence a 1-lattice with $k = 2$ is a ring, a 2-lattice with $k = 4$ is a two-dimensional square grid, an so on (see Figs. 2.3 and 2.4 for examples). In principle, k can be any number (although it makes sense to require $k \geq 2d$), and so we could have a 1-lattice with $k = 10$, in which case nearest neighbours, next-nearest neighbours, and so on would be connected (see Fig. 2.5 for another example). These structures are particularly convenient because their characteristic path lengths and clustering coefficients can be calculated explicitly. For a 1-lattice with even $k \geq 2$, simple enumeration shows

$$L = \frac{n(n+k-2)}{2k(n-1)}$$

and

$$\gamma = \frac{3}{4}\frac{(k-2)}{(k-1)}.$$

It is obvious from these statements that L for a 1-lattice scales linearly with respect to n (for large n) and inversely with respect to k. The same length-scaling property can be inferred by considering the distribution sequence of a 1-lattice. Again, simple enumeration leads to the conclusion that $|\Gamma^i(v)| = k$ for all v and i. Hence $\Lambda_j = jk$, which is linear in j—the number of "degrees of separation." Necessarily, a linearly increasing distribution sequence corresponds to linear length scaling with n. Unlike L,

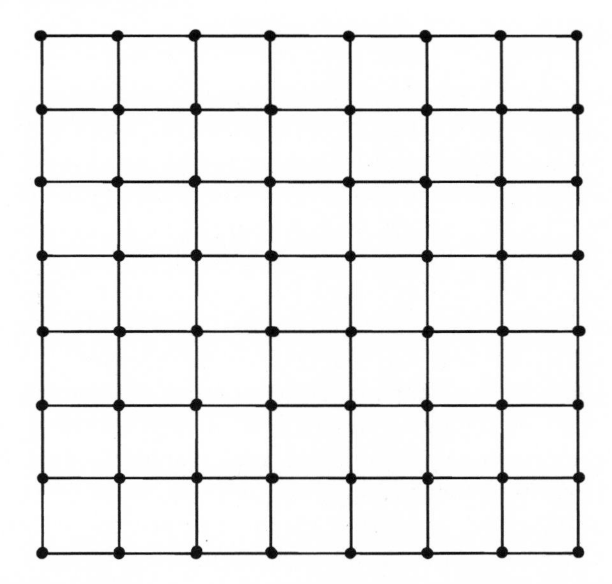

Figure 2.4 Example of a 2-lattice with $k = 4$ (boundaries are periodic).

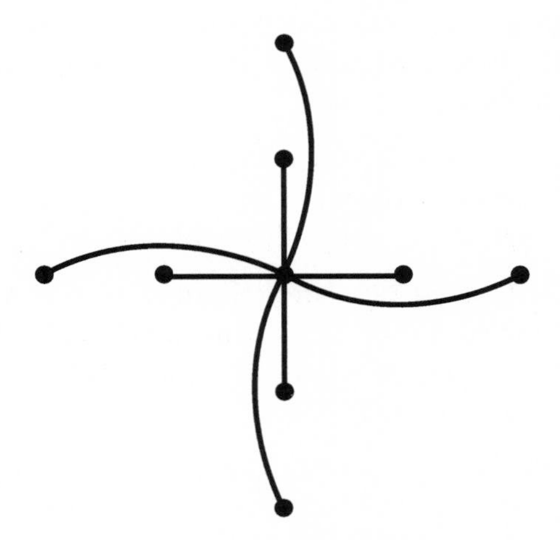

Figure 2.5 A single vertex in a 2-lattice with $k = 8$.

γ for a 1-lattice is independent of n and approaches 3/4 for sufficiently large k, at which point it is effectively independent of k also. *Hence any 1-lattice can be characterised by its length-scaling and clustering properties.* Similar statements hold for lattices in higher dimensions, which exhibit d-scaling as defined above.

Properties of Random Graphs

Strictly speaking, the kinds of random graphs discussed in this section do not appear very often in this book. However, they do constitute an important limiting case and, along with d-lattices, are frequently used as standard yardsticks in later chapters. They are also historically the root from which a number of serious investigations of the structure of social networks have sprung (the classic example being Harary 1959). Hence it is appropriate to run though at least the major classes of random graphs, some relevant terminology and definitions, and a few significant results that help in understanding and contextualising this work.

In its broadest sense, a random graph of order n is nothing more than a vertex set, consisting of n vertices, and an edge set that is generated in some random fashion. The set of all such graphs is called $\mathcal{G}^n$. Almost all of random graph theory, however, concerns itself with the analysis of one of two *models* of random graphs, referred to as $G(n, M)$ and $G(n, p)$, respectively, and the relationship between them. Many of the fundamental properties of these two models, along with the techniques used to analyse them, were developed in the late 1950s and early 1960s in a series of papers by Erdös and Rényi (1959, 1960, 1961a, 1961b), but all the material included here is referenced to the standard text on random graphs by Bollobás (1985).

Definition 2.2.13. $G(n, M)$ is a labelled graph with vertex set $V(G) = \{1, 2, \ldots, n\}$, having M randomly chosen edges (where M usually depends on n). $G(n, M)$ is frequently abbreviated as G_M.

Definition 2.2.14. $G(n, p)$ is a labelled graph with vertex set $V(G) = \{1, 2, \ldots, n\}$, in which every one of the possible $\binom{n}{2}$ edges exists with probability $0 < p < 1$, independent of any other edges. $G(n, p)$ is frequently abbreviated as G_p.

Random graph theory basically defines the conditions under which graphs belonging to either G_M or G_p possess a given *property* Q (for example, that it is connected), usually in the limit $n \to \infty$. Roughly speaking, broadly similar graphs with the same number of vertices share the same properties, and random graph theorists are interested in finding the conditions upon either M or p under which a particular property emerges in *almost all* graphs of the relevant model, in the infinite limit of n.

It turns out that for most purposes G_M and G_p are practically interchangeable, provided that $M \simeq pN$. G_p is easier to prove theorems with because the edges are independent, whereas in G_M (because the total number of edges is fixed) there is necessarily some dependence of an edge being chosen, based on previous choices. This dependence is small, however, and does not affect any of the important results, so hereafter both models will be referred to simply as random graphs.

One of the most striking results of random graph theory is that most monotone properties[8] appear suddenly. That is, there exists a *threshold* function $M^*(n)$ that determines whether or not a graph is either very unlikely or very likely to have the property Q. This threshold can be defined in a number of ways, but perhaps the most intuitive is to think of $\mathcal{G}^n$ as a *graph process*. That is, starting from a vertex set with no edges, edges are added one-by-one in a random fashion, where each addition is regarded as a unit of time. The threshold function $M^*(n)$ is then regarded as a critical time, before which the property is unlikely to exist, and after which it is very likely. There are certain technical issues surrounding the uniqueness of these functions and how they differ between models, but these are not of concern here. The important thing to understand is that if we imagine random graphs as dynamic "organisms," growing in time, then the appearance of practically any property of interest will occur on a timescale that is very short compared with the timescale of the whole process. Similarly, if we imagine the development of random graphs as a trip through parameter space, then all the action happens in a very narrow region of that space.

This *threshold function* is strongly reminiscent of second-order phase transitions that have been well studied in statistical physics (see, for example, Stauffer and Aharony 1992) and that appear in the dynamical systems in Part II. Furthermore, even though the graphs considered in this book are not random graphs in the strict sense, and even though the number of edges $M(n)$ is preserved for all parameter values, precisely this kind of rapid transition occurs in terms of their large-scale, statistical properties. This interesting similarity and its connection with random graph theory receive more attention in Chapter 4.

Of all properties Q, none seems to have received more attention than *connectedness*. At which point in a graph process do the graphs become connected? What is their structure before they become connected? How do they make this transition? And once they are connected, how connected are they? That is, how many edges could be removed before they would once again be disconnected, and what is the expected distribution of completely connected subgraphs (*cliques*)? These are some of the major issues that have been addressed by random graph theorists over the last forty-odd years. However, because the primary statistic of interest here is the characteristic path length (L), and because disconnected graphs have infinite L, only connected graphs will be considered. Admittedly this approach glosses over some important and interesting questions that are relevant to a completely general treatment of length in graphs. However, the resulting simplicity is useful for a first pass at the problem, and there is still much of interest to be learned. As far as random graphs are concerned, a famous theorem by Erdös and Rényi (1959) guarantees that "almost any" random graph with more than $n/2 \ln(n)$ edges (equivalent to $k \gtrsim \ln(n)$) will be connected.[9] In practice, for finite n, it is sufficient simply to set $k \gg 1$ and check that no disconnected graphs are generated.

A final issue of interest is that of the *diameter* D of a random graph. This is important because, in random graph theory, the diameter is the principal measure of the characteristic path length of a graph, presumably because it is easier to prove theorems about than measures like the mean or median shortest path length. As we will see in Chapter 4, however, the mean path length of a random graph is dominated by the diameter, and so all the important results about diameter apply, more or less, to the notion of characteristic path length. There are two results concerning the diameter that are of interest here: the diameter of a random graph with a maximal degree of two, and the diameter of a random graph of arbitrary degree. The reason for the distinction is that one of the models in Chapter 3, for reasons to do with connectivity, is based on a regular connected substrate of degree two. Hence the length-scaling characteristics of such substrates are important to know. For instance, a cycle (or topological ring) with n vertices and $k = 2$ has $D \simeq n/2$ and thus exhibits linear length scaling. It turns out that this is true for any graph with a maximal degree of two, which is really equivalent to saying that *any* regular graph with $k = 2$ is a ring and so will exhibit a characteristic path length that scales linearly with respect to n. As random graph theory is almost always concerned with the properties of random graphs as $n \to \infty$, little is said about D as a function of n for arbitrary k. Still, two things seem clear from Bollobás's treatment of the problem:

1. Almost all random graphs with the same n have the same D, for sufficiently large p (or k).
2. Random graphs are likely to be "spreading"; that is, the jth neighbourhood $\Gamma^j(v)$ includes almost as many "new" vertices as allowed by the finite degree of the graph. Hence the number of vertices within

a distance j of any vertex v is never much less than $k(k-1)^{j-1}$. This represents an exponentially growing distribution sequence, which implies that $j_{\max} \sim \ln(n)/\ln(k)$.

Both these statements will be useful in the later numerical experiments: the first because it implies that the details of the construction algorithms of Chapter 3 should not be important in the random limit, and the second because it means that random graphs must have close to the smallest possible L for any fixed n and k.

2.2.6 Dimension and Embedding of Graphs

Although graphs are not usually defined in terms of any underlying Euclidean space, and most problems in graph theory do not require graphs to exist in any such space, it will still be useful to think about what dimension would be required if they did. More specifically, any given graph can be thought of as a set of points embedded in a Euclidean space, where the Euclidean distance between any two points is just the shortest path length between the corresponding vertices, to within some distortion. The operative question then is, For a given distortion, what is the minimum dimension Euclidean space required to embed a given graph? Of course, there is nothing unique about the Euclidean metric, and we could just as well ask the same question of any metric space. But Euclidean spaces are a natural choice because they are familiar, and also because the limiting cases for some of the models in later chapters is a d-lattice, which embeds precisely into $\mathbb{R}^d$. This dimension is the *embedding dimension*, defined as follows.

Definition 2.2.15. For a graph G and distortion $c \geq 1$, the *embedding dimension* $\dim_c(G)$ is the least dimension d such that there is an embedding ϕ of G into $\mathbb{R}^d$ where every two vertices $i, j \in G$ satisfy

$$d(i, j) \geq \|\phi(i) - \phi(j)\| \geq \frac{1}{c}\, d(i, j).$$

A theorem by Linial et al. (1995) guarantees that any graph with n vertices can be embedded in $\mathbb{R}^{\dim_c}$ with distortion $c \leq (1+\epsilon)c^*$, where $\dim_c = O(\log n)$ and $c^* = O(\log n)$. Linial's theorem is rather more general than this, but this restricted version is sufficient.

At this point, a few observations seem appropriate. First, this theorem (which Linial et al. actually present as an embedding algorithm) is closely related to the techniques of *multidimensional scaling* described in Section 2.1.1.

Second, for a fixed n, graphs with different topologies will, in general, have different embedding dimensions. For instance, it is obvious (even without Linial's theorem) that a d-lattice of any size will always have an embedding dimension of d, whereas for a random graph, we are only guaranteed that $\dim_c(G) = O(\log n)$. The implication here is that, at least in the limit $n \to \infty$, random graphs "live" in $\mathbb{R}^\infty$, whereas in the same limit, d-lattices "live" in $\mathbb{R}^d$. This raises the conceptual challenge of what would happen to the dimension of a graph if it were to have its edges switched around one-by-one, causing it to move from a d-lattice to a random graph. This is a significant issue that will not be resolved here and that seems to be an interesting open research question.

Finally, if a graph has an embedding dimension d, then we might expect that its distribution sequence would grow like $\Lambda_j \propto j^d$. Certainly this seems plausible, if only by analogy to the distribution sequence of a d-lattice, which must necessarily grow in this fashion. The corresponding result for random graphs would be a distribution sequence that grows exponentially. The flip side of this observation is that one might expect any graph whose characteristic path length $L(n)$ scales logarithmically with respect to n to have a distribution sequence that grows exponentially with distance, and so it can be embedded only in a $\ln(n)$-dimensional space. One of the main results of the next chapter is that such graphs appear to be much more common than one might think.

2.2.7 Alternative Definition of Clustering Coefficient

Since the original publication of this book, an alternative definition of the clustering coefficient has been proposed (Newman et al. 2002) as follows:

Definition 2.2.16. Clustering coefficient for a graph G is $\gamma = \frac{\sum_v |\Gamma_v|}{\sum_v \binom{k_v}{2}}$,

where we have effectively reversed the order of operations from 2.2.11, putting summation before division. Whereas definition 2.2.11 quantifies the probability that two neighbours of a randomly chosen vertex will be connected, definition 2.2.16 quantifies the probability that two vertices possessing a mutually adjacent vertex will be connected. These quantities sound very similar, but in certain cases, they can yield quite different results. For example, 2.2.16 weights the contribution of high-degree vertices more heavily than low-degree vertices (whereas 2.2.11 weights all vertices equally), making it more appropriate for graphs that are characterized by the presence of a few very highly connected "hubs," a topic that is beyond the scope of this book, but has been widely discussed subsequently (Barabasi and Albert 1999).

3

Big Worlds and Small Worlds: Models of Graphs

One way to answer the question posed in Chapter 1—What are the most general conditions under which the elements of a large, sparsely connected network will be "close" to each other?—is to experiment with different *models of graphs*. Graphs are the appropriate constructions because, at this stage, the nature of the elements of the "system" is unimportant—all that matters is the fashion in which they are connected. At this stage we have some notions about the properties of two specific classes of graphs, namely d-lattices and random graphs. More specifically we have formulae for the length and clustering coefficient of a 1-lattice (with even degree k) and some arguments about the length and clustering properties of random graphs. In terms of their length characteristics, it is a plausible contention that these two classes are limiting cases in a "universe" of possible topologies, between completely correlated, one-dimensional structures and completely uncorrelated, high-dimensional structures (assuming, of course, that we are considering only connected structures). The plan, then, is to peruse the span of possible topologies that lie "between" these two limiting cases and see if anything interesting turns up along the way.

A number of different construction algorithms can be used to do this, but these appear to divide naturally into two categories: *relational graphs* and *spatial graphs*. Each of these categories consists essentially of a one-parameter family of graphs that interpolates between ordered and random extremes, where different models in the same category can be understood in terms of a single, model-independent parameter. Each value of the parameter defines an infinite set of graphs of varying n and k that are *topologically similar* in the sense that their length and clustering properties depend solely on n and k. Graphs chosen from different categories, however, appear to utilise entirely different mechanisms for traversing between the ordered and random limits. Relational graphs have the defining property that the rules governing their construction do not depend upon any external metric of distance between vertices. This is something of a fine point because the vertices of relational graphs are labelled and usually ordered according to some kind of geometry (such as a ring). But the distance between vertices is measured solely in terms of the graph itself, not in terms of any externally defined space. The converse is true for spatial graphs, which are explicitly embedded in some low-dimensional Euclidean space and whose construction depends explicitly on the resultant *spatial* distances between vertices. It turns out that this distinction is significant in that only relational graphs are capable of displaying the small-world phenomenon. It should be noted, however, that the spatial graphs considered here display a particularly strong spatial dependency, in that their distribution of edges tends to exhibit a finite *cutoff*—a fact that seems to play an important role in their other properties. It appears, in fact, that spatial graphs with more exotic distributions do display small-world features, but the matter is formally unresolved.

3.1 RELATIONAL GRAPHS

In the category of relational graphs, three successive models are presented, each of which represents a one-parameter family of graphs, spanning a broad range of structure from ordered through random graphs. The first of these (the α-model) is designed to represent the construction of a network in a fashion reminiscent of how a real social network is formed, that is, as a function of the currently existing network. The second model (the β-model) is an attempt to investigate the generality of the phenomena observed for the α-model, by removing any pretence of a social network, but spanning a similar range of graph structures. The third model (the ϕ-model) is motivated by a desire to unify the observed properties of the α- and β-models, in terms of a *model-independent* parameter (ϕ), as a function of which *all such models display the same characteristic transitions*. This result leads to a better understanding of the small world-phenomenon through the introduction of a class of *small-world graphs*: highly clustered graphs with small characteristic path lengths.

3.1.1 α-Graphs

The first (hence α) model represents a primitive attempt to capture the nature of connections in a *social* network. As discussed in Chapter 2, a number of social-network theorists have utilised the concept of a "social space" in which people exist as points and between which distances can be measured according to some appropriately defined metric. Unfortunately this approach runs immediately into trouble because of the inherent difficulty both of characterising the space (which is all but unknown) and defining the metric (equally so). The devil here is in the details because there is apparently no consistent and comprehensive method of characterising all human values, idiosyncrasies, and foibles that then aggregate to produce equally enigmatic human relationships—at least no method that is any less complicated and detailed than the phenomenon itself. This, of course, is always a central problem in any mathematical modelling assignment: to abstract the essence of the phenomenon to be described by stripping away detail without stripping away essential detail. If all the detail is essential, then either the problem is a hopeless one, immune to mathematical analysis, or else we are asking the wrong question. If, on the other hand, some fundamental principle is at work, then perhaps we have some hope of understanding it through a simple and comprehensible model. Optimistically adopting the latter frame of mind, let us make two generalisations that are, we hope, sustainable:

1. Assume that a network can be represented purely in terms of the connections between its elements: that whatever combination of factors makes people more or less likely to associate is accounted for and represented by the distribution of those associations *that actually form*. Hence we need not trouble ourselves (for the moment) with questions of spaces and metrics: only connections. Furthermore assume that all such connections are symmetric and of equal significance: that is, given some definition of what is required in order to "know" someone (whatever it may be), either two people know each other or they do not.
2. Assume that the likelihood of a new connection being created is determined in some fashion by the already-existing pattern of connections. In other words, a person's current friendship circles determine—to some (unknown) extent—their future acquaintances.

Exactly *how* existing relationships determine new ones is a big part of the mystery. We might imagine a world in which everyone has precisely one, completely connected group of acquaintances. In such a world, which we might term the *caveman* world, "Everybody you know knows everybody else you know and no one else." A less extreme example might substitute "most everyone," with the advantage that this kind of world could be connected (hence a *connected-caveman world*), whereas a caveman world would necessarily be disconnected into many isolated "caves." At the other extreme is a world in which we might imagine the influence of current friendships over new friendships to be so slight as to be indistinguishable from random chance. We might call this world *Solaria* after the planet of the same name in an Isaac Asimov novel (1957), where future humans live in isolation and interact, via robots and computers, as readily across the planet as they do with their spouses. We may be seeing the beginnings of such a world already in the proliferation of Internet "chat rooms" where complete strangers, with little in common, can meet, interact, and sometimes even end up marrying each other. Of course the real world, in which most of us live, is not like Solaria, nor is it caveman-like. In reality we have many different circles of friends, within each of which most people know each other, but between which relatively little interaction occurs. Another way to think about this is that any one *group* of people can be described approximately by a few representative attributes (for instance, university mathematics professors or actresses), but that any one person will defy such categorisations, spanning possibly quite disparate groups (a university mathematics professor who is also an actress).

The α-Model

So it seems that the real world of social networks lies somewhere between the "caveman" and "Solaria" extremes described above. Precisely where, of course, is anybody's guess. Hence rather than make some arbitrary assumption about the importance of existing connections versus random chance in the *real world*, a more successful approach might be to examine an *entire universe of possible "worlds"* and interpolate, by virtue of a single, tunable parameter, between its extremes. In order to work, this approach would require an algorithm for constructing graphs that explicitly embodies the following features:

1. At one extreme (the caveman world), the propensity of two unrelated people (which means that they share no mutual friends) to be connected is very small. Once they share *just one* friend in common, however, their propensity to be acquainted immediately

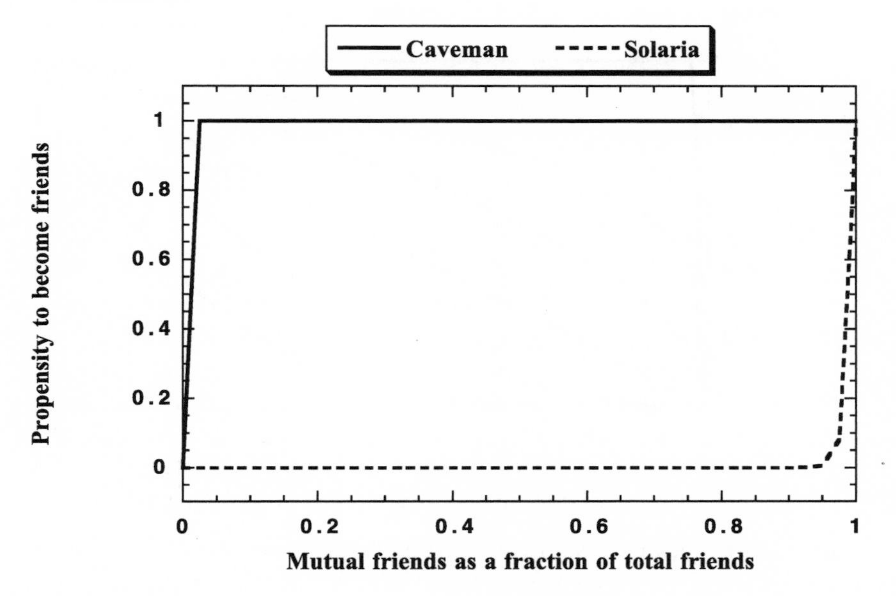

Figure 3.1 "Propensity to become friends" as a function of fraction of "mutual friends" in the two extreme cases considered for the α-model.

becomes very high and stays that way regardless of how many additional mutual friends they may have. So in worlds like this one, it is almost a certainty that the only people anyone will ever connect to are those with whom they share at least one mutual friend. So if we were to plot *propensity to become friends* against *fraction of current mutual friends*, it would start near zero, rise very rapidly to some relatively large number (which is normalised to one), and then plateau (see Fig. 3.1)

2. At the other (Solaria) extreme, no one has much propensity to connect to anyone in particular. In this sort of world, the propensity versus mutual friends curve would start near zero and stay near zero right up until the point at which all friends are mutual friends and then suddenly jump to one (see Fig. 3.1).[1]
3. In between these two extremes, the propensity curve could take any one of an infinite number of intermediate forms (see Fig. 3.2 for some examples), in which the only enforced condition is that the dependency be smooth and monotonically increasing with respect to increasing mutual friends.

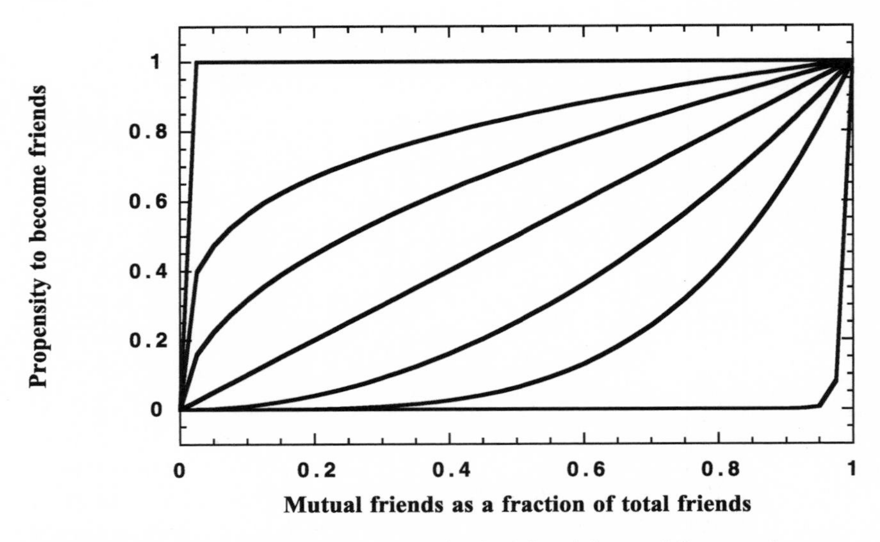

Figure 3.2 Between the two extremes, the functional form of "propensity to become friends" can take any one of an infinite number of intermediate forms. Some examples are shown here.

These conditions could be satisfied by a number of constructions. However, the following is as good as any:

$$R_{i,j} = \begin{cases} 1, & m_{i,j} \geq k \\ \left[\dfrac{m_{i,j}}{k}\right]^{\alpha}(1-p)+p, & k > m_{i,j} > 0 \\ p, & m_{i,j} = 0 \end{cases} \tag{3.1}$$

where:

$R_{i,j}$ = a measure of vertex i's propensity to connect to vertex j (zero if they are already connected)
$m_{i,j}$ = the number of vertices which are adjacent both to i and j
k = the average degree of the graph
p = a baseline, random probability of an edge (i, j) existing $(p \ll \binom{n}{2}^{-1})^2$
α = a tunable parameter, $0 \leq \alpha \leq \infty$.

This expression forms the basis for a construction algorithm that builds a graph of specified n, k (where k is the average degree), and α. The algorithm proceeds as follows:

1. Fix a vertex i.
2. For every other vertex j, compute $R_{i,j}$ according to Equation 3.1, with the additional constraint that $R_{i,j} = 0$ if i and j are already connected, and
3. Sum the $R_{i,j}$ over all j and normalise each to obtain variables $P_{i,j} = R_{i,j} / \sum_{l \neq i} R_{i,l}$. Then, since $\sum_j P_{i,j} = 1$, $P_{i,j}$ can be interpreted as the probability that i will connect to j. Furthermore, $P_{i,j}$ can be interpreted geometrically as follows: divide the unit interval (0, 1) into $n - 1$ half-open subintervals with length $P_{i,j}$, $\forall j \neq i$.
4. A uniform random variable is then generated on (0, 1). It must fall into one of the subintervals, say, the one corresponding to j_*.
5. Connect i to j_*.

This procedure is then repeated until the predetermined number of edges ($M = (k \cdot n)/2$) has been constructed. The vertices i are chosen in random order, but once a vertex has been allowed to "choose" a new neighbour, it may not choose again until all other vertices have taken their turn. Vertices may be "chosen," however, arbitrarily often, and this leads to a nonzero variance in the degree k (for all α). But the fact that all vertices are forced to make a new connection before any others are allowed to choose again ensures at least that no vertices will be isolated (as long as $k \geq 2$).

Having defined this construction, two questions arise. First, how do the statistical properties of characteristic path length (L) and average clustering coefficient (γ) depend upon α for fixed n and k? That is, given a vertex set (V) and a fixed number of edges ($M = (k \cdot n)/2$), how does the *arrangement* of those edges (according to Equation 3.1 for some fixed α) affect the resulting graph's properties? Second, for a given α, how do L and γ vary as functions of n and k? That is, how do graphs with fixed α *scale* with n and k, and do these scaling laws change for different α? Before attempting to answer these questions, it is necessary to understand more about the limiting cases of $\alpha = 0$ and $\alpha \to \infty$ to make sure that the model does indeed interpolate between reasonable and meaningful extremes.

The $\alpha = 0$ Case: Problems with Connectedness

When $\alpha = 0$, new connections are determined almost exclusively by the arrangement of existing connections. If $m_{i,j} = 0$, then $R_{i,j} = p \ll \binom{n}{2}^{-1}$, which is obviously very small. By contrast, if $m_{i,j} > 0$, however small, then $R_{i,j} = 1$. This means that the unit interval into which all the $R_{i,j}$ are normalised is occupied almost entirely by subintervals that represent those vertices with which the vertex i already has mutual connections. Consequently, given an existing subgraph of "friends," it is clear that new "friendships" are overwhelmingly likely to be added that increasingly interconnect that subgraph without extending it much into new territory. If the existing network is disconnected (partitioned into connected subgraphs), then it is highly likely to remain that way, and friendship circles will become denser without extending themselves at all (hence the caveman world). This result is only "highly likely" (rather than certain) because a random connection joining two subgraphs, however unlikely, is always possible (hence the small but nonzero p). If the existing network is connected, then precisely *how* it is connected will determine how much existing subgraphs fill themselves in and how much they "spread."

The Need for a Substrate

It should be clear by now that there is a problem in all this with the initial conditions. If existing edges predominate in the selection of future edges, then what happens when there *are* no existing edges? One might imagine that edges will at first form, randomly until some connected subgraphs form and that these will then provide the basis for future edges. Indeed, this is basically what happens. Unfortunately, a side effect of allowing clusters to form randomly is that, for small α, they remain disconnected from one another. Of course, as k is increased, these clusters become larger, and for sufficiently large k they will connect. However, numerical experiments show that the average fraction of pairs that can be reached from one another for $\alpha = 0$ grows only linearly in k (see Fig. 3.3), and thus the graph would become connected only at the cost of violating the sparseness condition ($k \ll n$). Hence small-α graphs necessarily are disconnected and so have infinite L, which makes comparison with connected graphs difficult.

Nevertheless, it is interesting to see what happens as α increases. Figure 3.4 shows the characteristic length of the connected *components* of α-graphs. For small α, the graphs are indeed disconnected, and so $L(\alpha)$ is effectively the average length of many small components. As α increases, the clusters gradually connect via randomly assigned edges, and the characteristic component length increases to a maximum, at which point the entire graph is connected. After this, $L(\alpha)$ drops rapidly to its

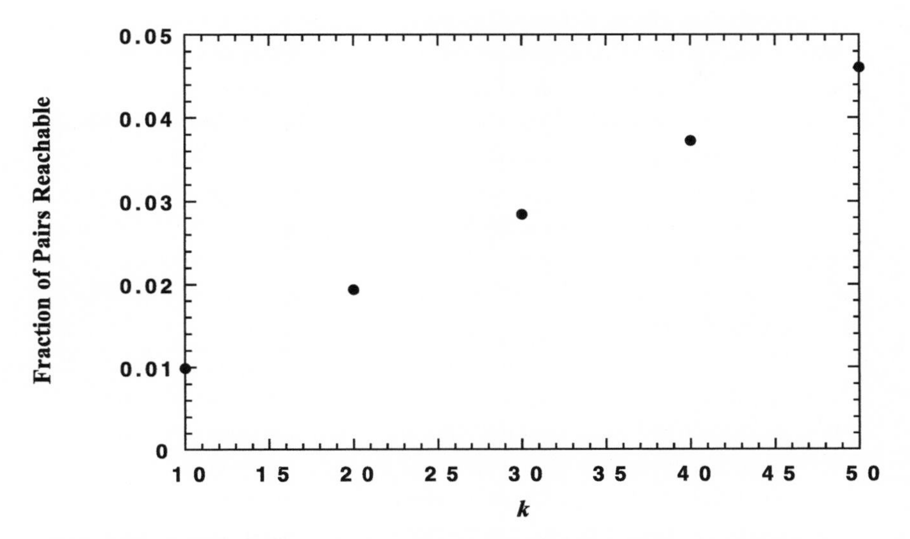

Figure 3.3 Fraction of all pairs of vertices that are connected vs. k for α-graphs with no substrate ($n = 1{,}000$, $k = 10$).

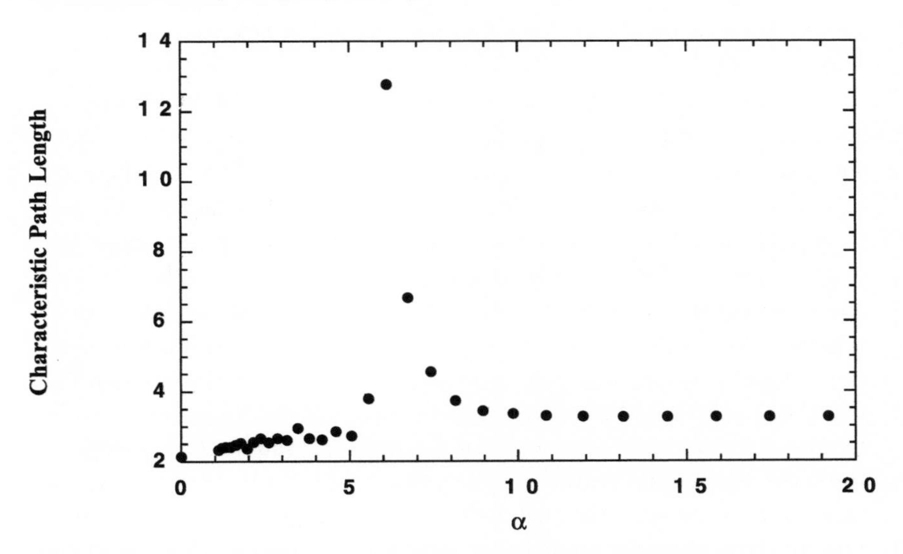

Figure 3.4 Average characteristic path length of connected clusters vs. α for α-graphs ($n = 1{,}000$, $k = 10$). Note that L rises to a maximum, at which point the entire graph becomes connected, and then decreases with increasing α.

asymptotic random graph limit as the edge assignments become less and less correlated with previously constructed edges.

The question of connectedness is an interesting one, and the phenomenon of small, connected components being joined together into a single, connected whole is reminiscient both of the threshhold function of connectivity in random graphs and of the percolation threshold from statistical physics. The mechanism at work here—increasing randomness of a fixed number of edges, rather than an increasing number of random edges or an increasing number of randomly occupied sites—is different, but the similarity with random graphs and percolation is intriguing nonetheless. For the problem at hand, however, the most obvious question raised by Figure 3.4 is How can we characterise the graphs to the left of the maximum, that is, the disconnected graphs? Clearly they cannot be regarded as having smaller characteristic lengths than connected graphs, as is suggested by the measure of average component length, but neither is it helpful to reject them simply as having infinite lengths that are meaningless to consider. It is obviously possible to add edges, after the fact, specifically for the purpose of connecting the graph. But this seems too arbitrary. It would also be possible to increase the amount of randomness in the construction (perhaps by increasing p), but this subverts the original aim of investigating the effects of increasing disorder. The question really is How can we introduce sufficient structure into the α-model such that all α-graphs will be connected without decorrelating the edges and without introducing any *more* structure than is necessary?

One way to answer this is to build into the graph-construction procedure an initial, connected substrate upon which all future edges are added. This is a significant step, and it raises the concern that perhaps some nongeneric properties will be built in as a result. In fact, it turns out that quite different things happen in the currently disconnected region of the α-model, depending on which substrate is chosen, but that certain generic features of the model emerge despite this. For now, let us proceed cautiously to search for a connected substrate that imposes the least restrictions on the resulting family of graphs, that is, one that exhibits the following characteristics: (1) It must exhibit *minimal structure*, in that no vertices are to be identified as special. This eliminates structures like stars, trees, and paths that have centres, roots, and end points, respectively, (2) It must be *minimally connected*; that is, it contains no more edges than necessary to connect the graph in a manner consistent with the minimal structure requirement.

The only structure that satisfies both these criteria is a topological ring, that is, a graph in which every vertex has precisely two edges, none of which are duplicated. Necessarily, the deletion of any two edges would cause the graph to become disconnected. It might be argued that the very homogeneity of the ring is, in itself, a considerable degree of structure—indeed this would be true if it were the finished product. However, the point of the ring is that it is only an initial condition upon which the construction algorithm acts. As we shall see, the nature of the final graph depends overwhelmingly on the value of α in Equation 3.1. Furthermore, because the ring is (after a chain) the graph with the maximal characteristic length for a given n and k, it allows for the greatest possible range of outcomes in terms of the length of the final graph. For instance, if we were to start with a star (in which a single vertex—the centre—is connected to all other vertices, each of which are connected solely to the centre), then its characteristic length is already $L \approx 2$, and the addition of extra edges can make little difference. Such a substrate would clearly impose major restrictions upon the range of graphs that could be generated by the α-model. A ring substrate, however, allows everything from a highly clustered graph, in which the ring structure is largely preserved, to a very uncorrelated structure. Nevertheless, the introduction of any arbitrary structure should be treated with caution.

The $\alpha \to \infty$ Case versus $G(n, M)$

The first point to settle is whether or not, for sufficiently large α, the model converges to an adequate approximation of a random graph, unhindered by the presence of a deterministic ring structure embedded in it as a substrate. Certainly, as α becomes very large, the relative probabilities $P_{i,j}$ will all tend to the uniform random limit of $P_{i,j} \approx p$, and so, given that the number of edges is fixed at $M = (k \cdot n)/2$, it is at least plausible that the resulting α-graphs would resemble the random graph model described in Chapter 2. Numerically this seems to be true, and the large-α (Solaria) limit will generally be referred to as the *random limit*. Strictly speaking it differs from a random graph in two important respects, but neither appears to be critical.

First, *the ring substrate persists even in the large-α limit.* This might seem to be a major objection to the inclusion of a particular substrate: such features do not necessarily occur in random graphs (although they probably do for sufficiently large n and k), and so α-graphs cannot have the same statistical properties as random graphs. In fact, it seems that any such substrate is "forgotten" after the addition of sufficiently many random edges such that all graphs with the same n, k, α have the same statistics regardless of what substrate they were initiated with.[3]

Second, *the construction algorithm does not choose edges in an entirely independent fashion.* Rather, it systematically loops over all vertices choosing one new neighbour for each vertex in turn. This is more restrictive than either the $G(n, p)$ or even $G(n, M)$ models, as it introduces greater correlations among the edges. However, the model is not too restrictive (not even as restrictive as an r-regular random graph) as vertices can be connected *to* at any time, thus allowing a nonzero variance in k.

Hence it seems reasonable to assume that, at least in the regime of $k \gg 1$, the α-model reduces effectively to a random graph when $\alpha \gg 1$. Given the extra correlations involved, it would probably be very hard to prove this equivalence rigorously. Hence numerical evidence in accordance with the expecations of random graph theory will have to do. Adopting this approach, then at what value of α could one expect the random limit to be approached? Roughly speaking, Equation 3.1 suggests that when $(m_{i,j}/k)^\alpha \ll p$, the algorithm will draw no distinction between edges that connect mutual friends and edges that connect vertices at random. For $k = 10$ and $p = 10^{-10}$ (the parameters used here), this reasoning suggests that the random limit would be approached for $\alpha \gtrsim 11$. Both Figures 3.5 and 3.11 support this estimate.

Length Properties

For the class of graphs generated by the α-model, no closed-form expression exists that yields the characteristic path length for arbitrary n, k, and α. For small-α, we might expect that the resulting length would be well approximated by a 1-lattice, although even here there turn out to be some differences associated with the breakdown in the regularity of the degree. At the other extreme, random graphs are helpful approximations, but again it is unlikely that any precise, analytical statements can be made about the expected length. The same is not true of numerical simulations, which can be used to compute the characteristic length exactly for small graphs (up to $n \approx 1{,}000$) and approximately for larger graphs (up to $n \approx 20{,}000$). For larger graphs still, even numerical methods run into problems, and (in the absence of more powerful computing resources) the heuristic models of Chapter 4 become necessary.

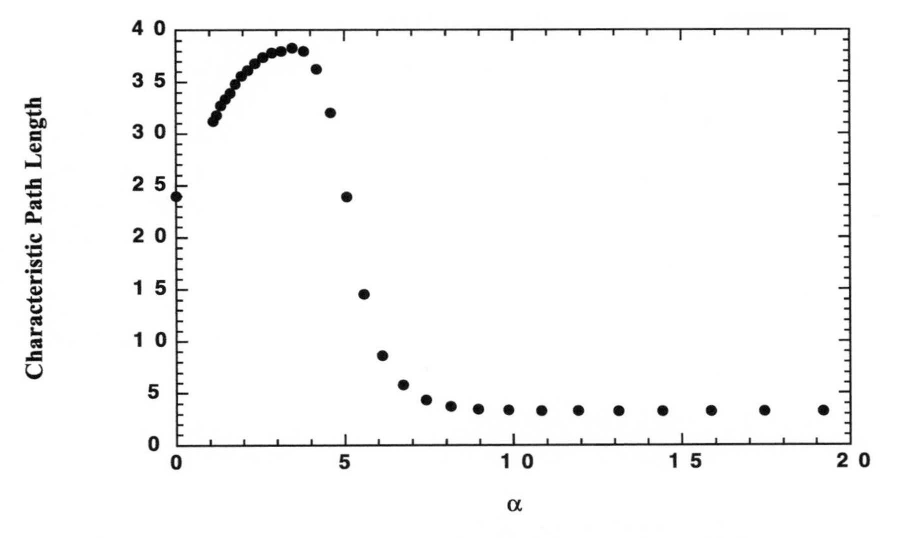

Figure 3.5 L vs. α for α-graphs constructed on a ring substrate and averaged over one hundred realisations of the construction algorithm ($n = 1{,}000$, $k = 10$).

Phase Transition

Having established that the α-model does indeed interpolate between defensible limits and that numerical simulation is the only feasible strategy for determining the characteristic path length as a function of n, k, and α, it is time to examine some results. Figure 3.5 shows $L(\alpha)$ plotted against α for a typical case of $n = 1{,}000$, $k = 10$, which satisfies the constraints that $n \gg 1$ and $n \gg k \gg 1$.[4] Three features of this picture are worthy of note:

1. For low values of α ($\alpha \lesssim 1$), $L(n, k, \alpha)$ starts at roughly two-thirds the value of an equivalent 1-lattice (with the same n and k) and then rises in a distinct hump. The reason for the hump appears to be that, for $\alpha = 0$, no distinction is made by the construction algorithm between pairs of vertices with one and more than one mutually adjacent vertices, and so a graph that is locally less redundant than a 1-lattice (and also less regular) develops. However, as α increases, but before $(m_{i,j}/k)^{\alpha} \ll p$, edges connect preferentially between vertices that have $m_{i,j} > 1$, and this results in weakly connected clusters much like the necklace of highly-interconnected clusters described earlier as the connected-caveman graph. Because these have close to the maximum length possible for any graph with

n vertices and $M = (k \cdot n)/2$ edges (where the variance in degree (k) of the vertices is not too great), graphs in this interval may be thought of as "big."
2. For high values of α ($\alpha \gtrsim 11$), $L(n, k, \alpha)$ is much smaller, and for α increasing much past 10, there is no appreciable dependence of L upon α at all. This asymptotic limit to $L(n, k, \alpha)$ is approximately that of a random graph, and, because a random graph is close to the lower limit of characteristic path length for a given n and k, graphs in this interval may be thought of as "small."
3. For intermediate α, there is a rapid but apparently smooth transition from "big" to "small" graphs.

Given the limits built into this model, some sort of transition from "big" to "small" graphs must occur, but it is not obvious that this transition should be as sudden and rapid as it appears to be. This behaviour is striking and reminiscent of the phase transitions that occur in magnetic spin systems in low dimensions (Palmer 1989) or in the characteristic size of connected clusters in low-dimensional percolation models (Stauffer and Aharony 1992). The difference is that the quantity being measured is not a thermodynamic property of the configuration or the onset of new structure as a result of the additional connections. Rather it is a property that is purely topological in nature: the total number of edges is conserved for all α, and only their arrangement is allowed to change. Dramatic though such a phase transition seems for n as small as 1,000 (over an order of magnitude), it would be of even greater significance if the size of the "cliff" were to increase with increasing n. In other words, do the graphs in the three regions exhibit different length *scaling* properties with respect to n and k?

Length-Scaling Properties

A rough characterisation of length-scaling properties will suffice here as the issue will be dealt with more carefully in Section 3.1.3 and again in Chapter 4. For now, the following scaling properties in the three regions of Figure 3.5 are noteworthy:

1. In the small-α region, $L(n, k, \alpha)$ scales linearly with n (Fig. 3.6) and with the inverse of k (Fig. 3.7). This is not surprising, as the length of a 1-lattice scales linearly with n (and inversely with k), and the graphs in this region appear closely related to 1-lattices.

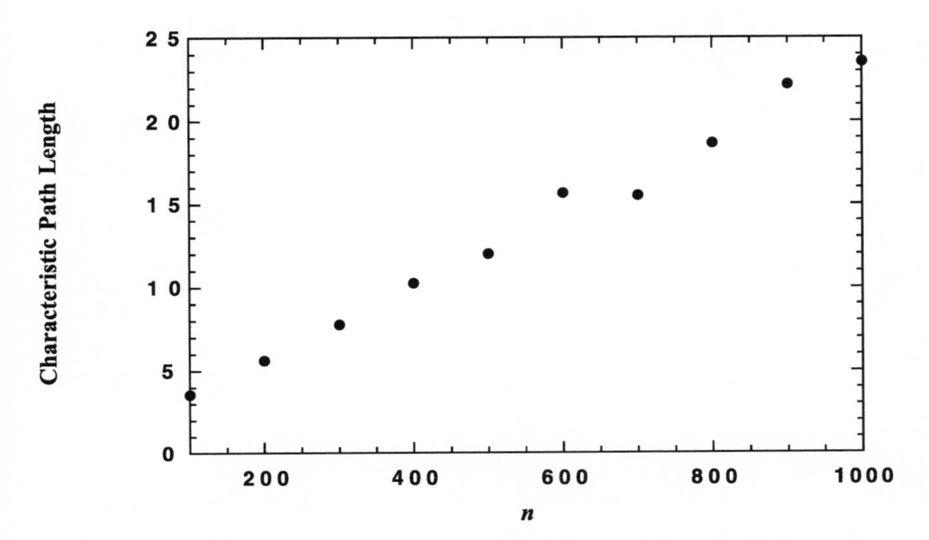

Figure 3.6 L vs. n for α-graphs with $\alpha = 0$, $k = 10$. L scales linearly with respect to n.

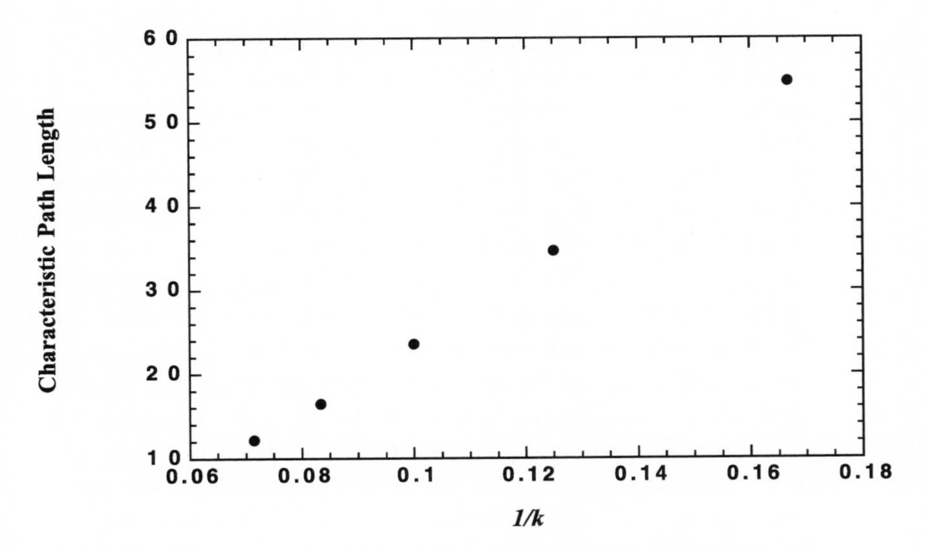

Figure 3.7 L vs. $1/k$ for α-graphs with $\alpha = 0$, $n = 1{,}000$. L scales inversely with respect to k.

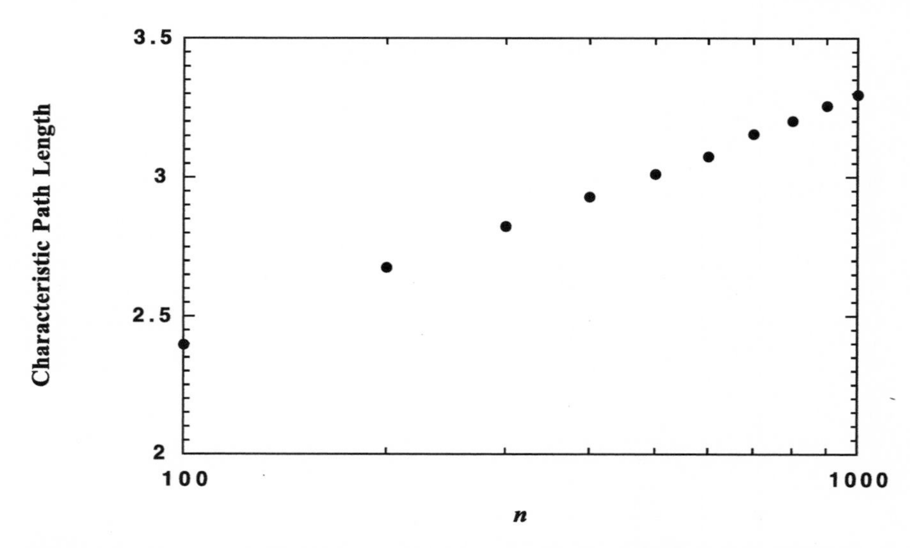

Figure 3.8 L vs. n for α-graphs with $\alpha = 10$, $k = 10$. L scales logarithmically with respect to n.

2. In the large-α region, $L(n, k, \alpha)$ scales logarithmically with n (Fig. 3.8). Once again, this is not surprising, as we know that the length of a random graph must scale logarithmically with n, and all graphs in this region appear to be well described by random graphs. Furthermore, whilst no formulae exist for L of a random graph, a reasonable *asymptotic* approximation (that is, in the limit of very large n and k) is $L_{\mathrm{random}} \sim \ln(n)/\ln(k)$ (see Chapter 4 for some details), so one would expect that large-α graphs would display $L \propto 1/\ln(k)$, as displayed in Figure 3.9.
3. What is surprising is that the graphs in the transitional region also appear to display logarithmic length scaling with n (Figure 3.10). The relationship is not very clear from the data, and the reason for this will be easier to understand after the discussion in Section 3.1.3. However, there definitely seems to be some change, not only in the length of α-graphs, once they enter the "cliff" area, but also in their scaling properties. This second change is perhaps even more significant than the first, because it implies that, as n becomes very large (some networks may have millions or even billions of elements), the difference between the lengths of graphs in "big" and "small" regimes becomes profound.

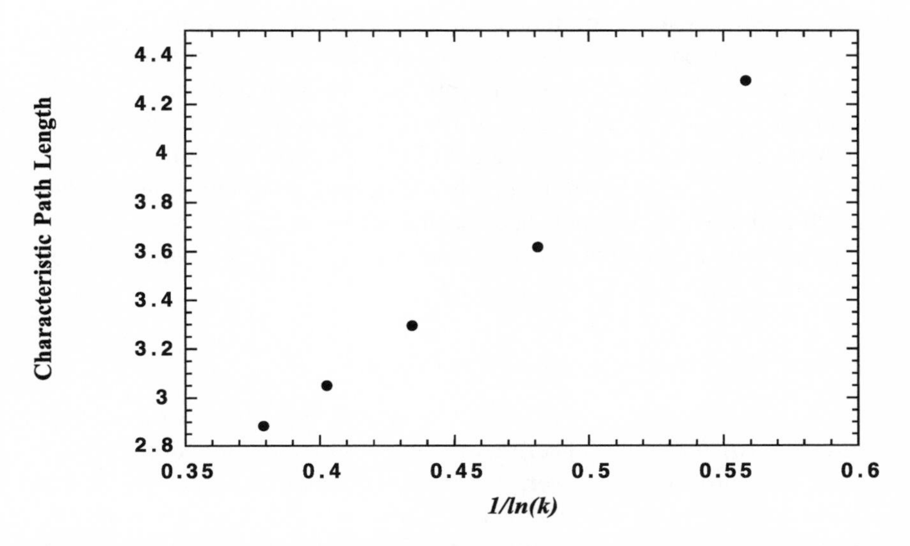

Figure 3.9 L vs. $1/\ln(k)$ for α-graphs with $\alpha = 10$, $n = 1{,}000$. L scales inversely with respect to $\ln(k)$.

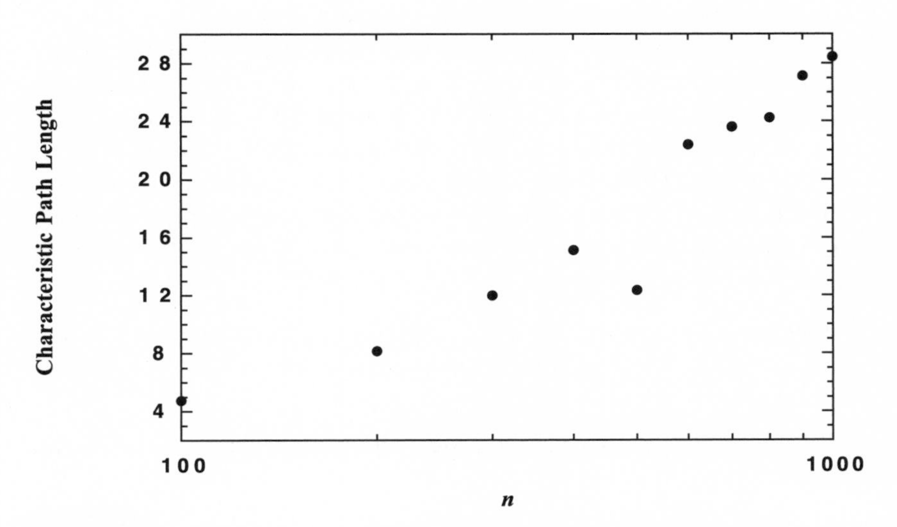

Figure 3.10 L vs. n for α-graphs with $\alpha = 4.5$, $k = 10$. L scales logarithmically with respect to n.

Clustering Properties

The second statistic to be measured as a function of α is the clustering coefficient, which, we recall from Chapter 2, is a measure of how densely connected (or "cliquish") local neighbourhoods are on average. That is, $\gamma(\alpha)$ is a measure of the degree to which vertices in the neighbourhood $\Gamma(v)$ are connected to other vertices in $\Gamma(v)$, averaged over all $v \in V(G)$. As with $L(\alpha)$, the $\gamma(\alpha)$ function should approach certain limits in the small- and large-α extremes. Specifically, $\gamma(0)$ should be close to γ for a 1-lattice, and, at the other extreme, $\lim_{\alpha \to \infty} \gamma(\alpha) \sim (k/n)$, which is what one would expect for a random graph.[5]

Another Phase Transition: Small-World Graphs

Figure 3.11 shows $\gamma(\alpha)$ for the same parameters as before ($n = 1{,}000$ and $k = 10$). At first glance, this picture looks very similar to that of $L(\alpha)$ in Figure 3.5. However, closer inspection reveals that whilst the broad features of $\gamma(\alpha)$ are similar to those of $L(\alpha)$, they are *shifted* in α, such that the "clustering cliff" occurs distinctly *after* the "length cliff." What is more, this appears to be true for all n and k.

The difference in the phase transitions can be made more obvious by normalising both quantities according to their respective values at $\alpha = 0$ and superposing them. The result, shown in Figure 3.12, shows clearly that for some interval of α, α-graphs exist that are as highly clustered as the most clustered objects that can be constructed with this model, yet have characteristic lengths that are of the same order of magnitude as random graphs. Furthermore, as all these graphs are on or to the right of the length cliff, they also exhibit logarithmic length scaling, implying that this result holds for all n. This observation leads to the following claim.

Conjecture. *There exists a class of graphs that are highly clustered yet have characteristic length and length-scaling properties equivalent to random graphs. These are called* small-world graphs.

Precisely how such objects can arise—and how special they are—is the subject of the rest of Part I.

More General Substrates: A Justification of Assumptions

The first thing that should be established is that the above conjecture is not dependent upon the inclusion of a substrate to initiate the con-

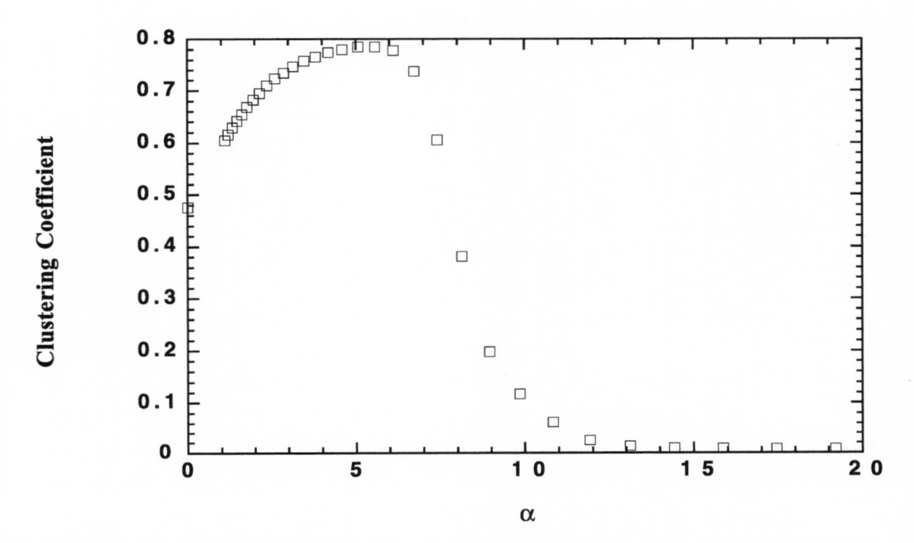

Figure 3.11 Clustering coefficient γ vs. α for α-graphs on a ring substrate, again averaged over one hundred realisations ($n = 1{,}000$, $k = 10$).

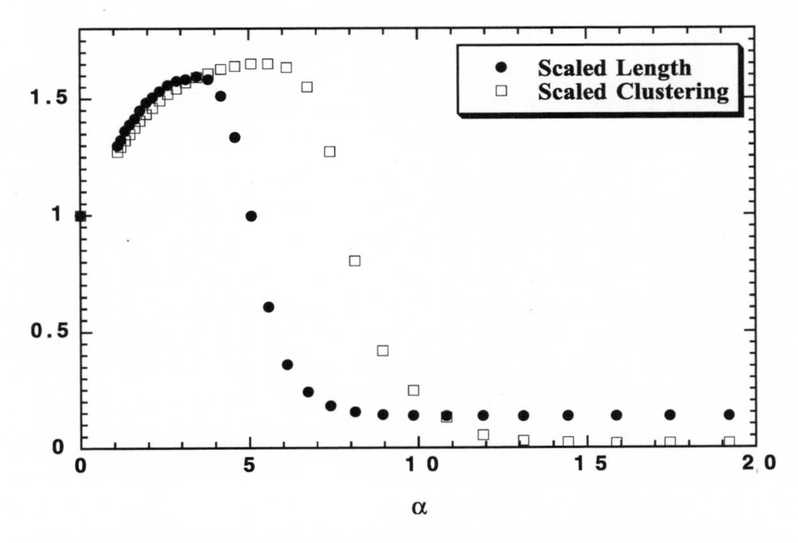

Figure 3.12 Scaled $\gamma(\alpha)$ and $L(\alpha)$ vs. α for α-graphs on a ring substrate ($n = 1{,}000$, $k = 10$).

struction algorithm. Recall that the justification for using a ring substrate is that it ensures the connectedness of sparse graphs with low α, whilst introducing a minimum of structure. Nevertheless, if it were to turn out that there was something nongeneric about the use of a one-dimensional ring as a substrate upon which subsequent edges were added—that all the properties observed so far are sensitive to this artificial construction—then very little could be concluded, because there is certainly no evidence to support the fact that the social (or any) world is founded on a ring. It is important then to sort this out before proceeding any further and committing ourselves to believing that the results so far are meaningful in terms of the original question posed.

Lattice Substrates. One obvious replacement of a ring substrate is simply a periodic, nearest-neighbour lattice of arbitrary dimension d. Simple reasoning implies that such a structure will exhibit length scaling proportional to $n^{1/d}$. The easiest way to see this is to imagine the graph as a nearest-neighbour lattice in a d-dimensional Euclidean space and realise that the volume of a d-cube in such a space grows like l^d, where l is characteristic length of one side of the cube. If n vertices are contained within the d-cube, then $n \propto$ volume, and hence $l \propto n^{1/d}$. The interesting question is whether or not this model, which has qualitatively different scaling properties from the ring-based model in the low-α limit, displays the same logarithmic scaling in the large-α limit. If so, over what range of α does the transition from $n^{1/d}$ scaling to logarithmic scaling occur?[6] In the case of $d = 2$, not only does the appropriate transition occur, but the cliff occurs in approximately the same place as for the ring substrate, and the qualitative nature of the transition looks the same (see Fig. 3.13). Only the height of the cliff differs, as a result of the fact that the upper plateau scales like $\sqrt{n}$ instead of n. This observation that both the limiting state of the graph and the qualitative nature of the transition between the small and large α states appear to be independent of the substrate is particularly significant.

Tree Substrates. When used as a substrate, a *Cayley tree* (see Fig. 3.14) places more restrictions upon the α-model construction algorithm than does the ring, because trees have definite roots and branches that distinguish some vertices as more central than others and some edges as more significant (in that their deletion would result in larger subgraphs becoming disconnected). Also, inspection of a tree growth structure reveals that the distribution sequence grows exponentially; hence the characteristic

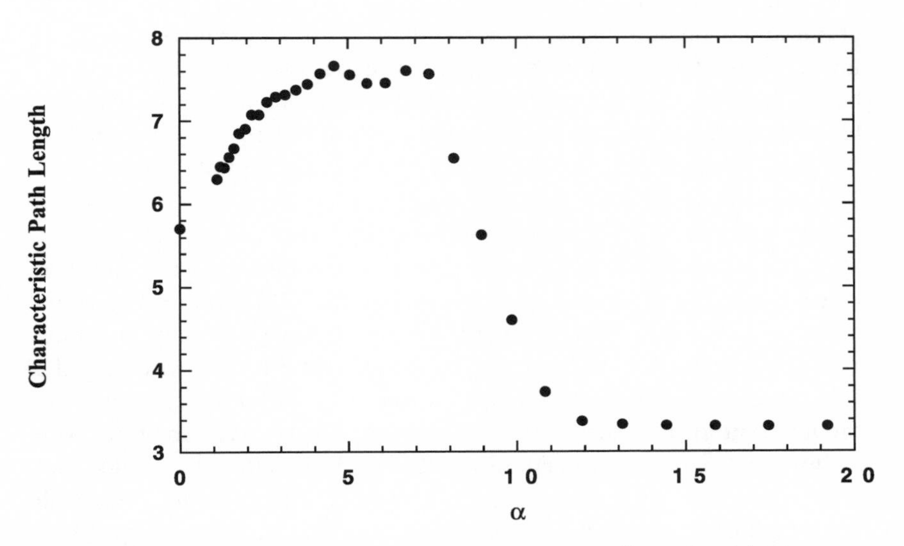

Figure 3.13 L vs. α for α-graphs constructed on a two-dimensional lattice substrate ($n = 1{,}024$, $k = 10$).

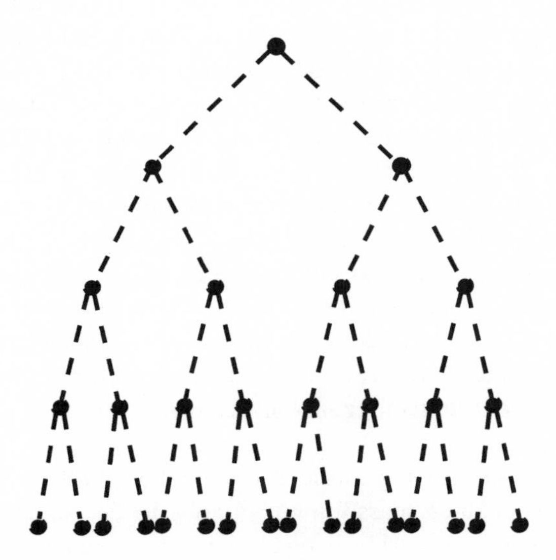

Figure 3.14 Cayley Tree substrate with branching of 2.

length of a tree must be logarithmic in n. This does not leave nearly as much "room" (in terms of characteristic path length) between a tree graph and a random graph, so a tree-based α-model cannot explore as much interesting terrain as a ring-based model can. Nevertheless, tree-based graphs are frequently encountered in problems to do with the lengths of graphs and have substantial applications to social systems, in the guise of organisational hierarchies, so it is appropriate only to consider what properties the α-model might display when constructed upon a tree substrate. The minimal branching for a tree is when each vertex after the root has one edge leading back to the previous layer and two edges branching out to the next layer (see Fig. 3.15). The results, again for $n = 1{,}000$, $k = 10$ (Fig. 3.16), appear to have precisely the same qualitative features as those for the 1-lattice and 2-lattice, with only the vertical scale changed. As anticipated, the scale difference results from the inherent logarithmic length scaling of the Cayley tree substrate. even without any random connections at all (Fig. 3.17). However, notice that once α becomes sufficiently large, it is no longer possible to distinguish the length of an α-graph built on a tree from that built on a ring.

Minimally Connected Random Substrates. At first one might think that the most general and so most defensible substrate on which our α-graphs can be built would be randomly connected. That is, just add edges at random until the whole graph becomes connected (recall from Chapter 2 that this should happen around $n/2\ln(n)$ edges) and then use this as the initial condition upon which edges are added according to Equation 3.1. But this approach has a problem, which is that a randomly connected graph is a *random* graph, and even the addition of highly correlated edges upon a random substrate will not negate its random nature. Hence a random substrate is not at all general in that graphs built on such a substrate are necessarily constrained to a narrow range of possible topologies. Figure 3.18 confirms this, showing that for α-graphs with a random substrate, a transition in $L(\alpha)$ still occurs, but of small magnitude and with no attendant change in scaling.

No Substrate at All. Having considered a number of alternate substrates, each of which enables sparse, low α-graphs to be connected, it is instructive to compare the various results with those generated for *no* substrate, considering only the values of α for which the resulting graphs are connected. A comparison of this parameter range reveals that connected *no*-substrate α-graphs show a greater statistical resemblance to

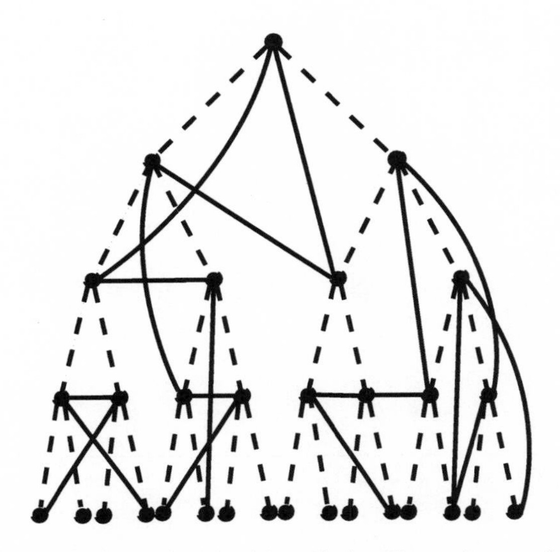

Figure 3.15 A possible α-graph with a Cayley Tree substrate for $\alpha = 0$ (substrate edges dashed).

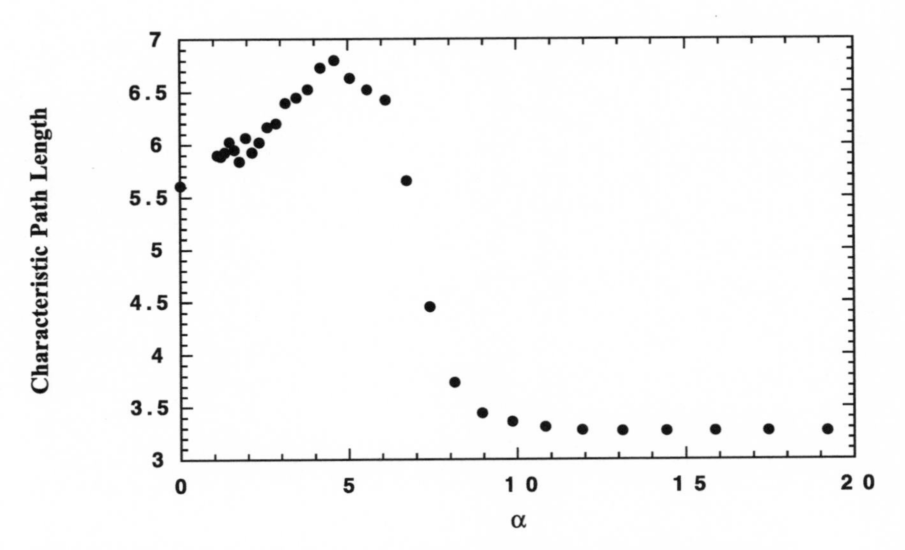

Figure 3.16 L vs. α for α-graphs constructed on a Cayley Tree substrate ($n = 1{,}000$, $k = 10$).

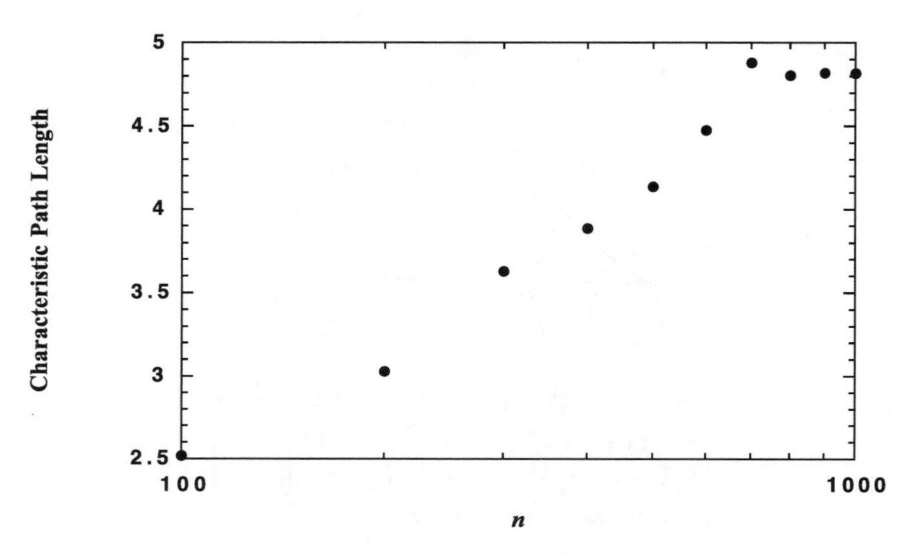

Figure 3.17 L vs. n for α-graphs on a Cayley Tree substrate ($\alpha = 0, k = 10$). Displays logarithmic scaling of L with respect to n.

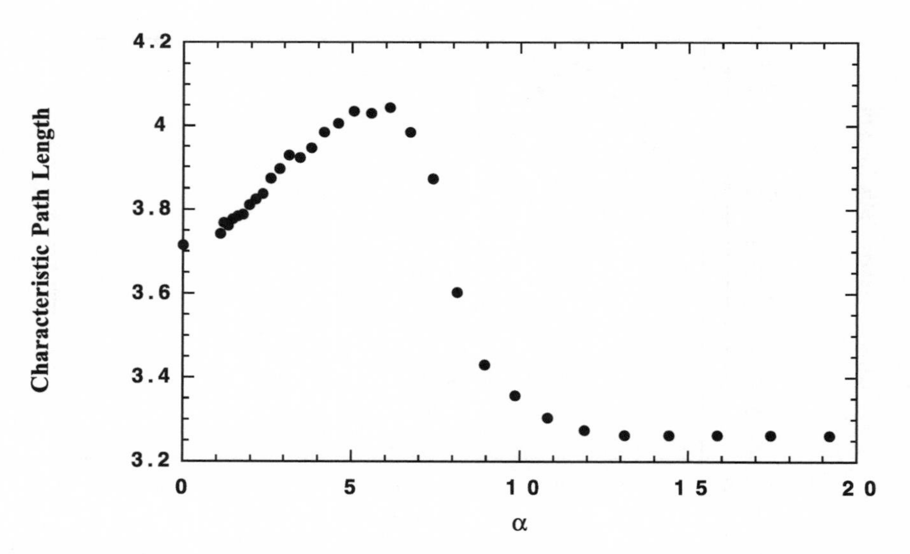

Figure 3.18 L vs. α for α-graphs constructed on a minimally connected, random substrate ($n = 1{,}000$, $k = 10$).

the ring-substrate graphs than to those generated on any other substrate (see Fig. 3.4). The agreement is still not complete over the entire connected range, but, as we shall see in Section 3.1.3, this discrepancy can be reduced by recasting the α-model in terms of the appropriate model-independent parameter. What this last result suggests is that, at the point where the *no*-substrate α-graphs become connected, they do so in a fashion resembling a minimally connected structure, that is, a ring.

Justification for Concentrating on Ring Substrates

Thus the justification for concentrating on α-graphs with ring substrates is more substantial than mere convenience and simplicity:

1. It is important to construct graphs that are connected for all values of α; otherwise, comparisons of characteristic path length would be meaningless. As this property does not emerge naturally from the α-model, then *some* substrate is required (some models that *are* connected naturally will be presented later, but these are no less artificial).
2. All substrates considered yielded qualitatively similar $L(\alpha)$ curves, indicating that the transition of the ring substrate from "large" to "small" is, at least in this respect, generic.
3. Of all substrates considered, the ring substrate allows for the greatest variation in $L(\alpha)$ over the full range of α, thus satisfying the main goal of exploring a wide range of possible topologies.
4. Of all the substrates considered, the ring substrate provides the best match with the *no*-substrate statistics over the range in which the *no*-substrate model yields connected graphs.
5. For sufficiently high α, all differences between substrates are erased, and the statistics produced by the various models yield indistinguishable results. The fact that this happens at lower α for L than for γ is not immediately obvious and therefore stands as a feature to be explained.

A Mystery?

The main point to be understood so far is the conjecture that graphs can be as highly clustered as those with very large characteristic path lengths yet have very short characteristic path lengths—almost as short, in fact, as is possible. Equally important is that *this observation does not appear to depend on the choice of substrate*. Hence the existence of these *small-world graphs* seems at once undeniable and yet still something of a mystery. Specifically, it is mysterious why the statistics of α-graphs seem to change so suddenly and rapidly as a function of α, why these transitions occur at different values of α for the different statistics (L and γ), and what changes in scaling laws (if any) attend these transitions. Where, in other words, do all these features come from? Are they somehow related to the construction methodology of the α-model? Is there something specific in the details of Equation 3.1 that generates small-world graphs, or is there some general mechanism at work here that is entirely unrelated to either the choice of substrate or model? It is not obvious how such a question should be approached, especially with only numerical data in hand, but one approach is to start out with a different, simpler, model and determine whether or not the same phenomena arise there.

3.1.2 A Stripped-Down Model: β-Graphs

This simpler, second model is motivated by the following factors:

1. The properties of α-graphs near the $\alpha = 0$ extreme are dominated by the properties of the substrate upon which they are built.
2. A ring substrate, by virtue of its simplicity, is also of considerable interest in its own right, especially in applications of coupled systems in the physical and biological sciences.
3. Whilst a ring substrate is not entirely generic, it *is* generic in the sense that the clustering behaviour is similar for all substrates, and that for α not too small, the lengths of α-graphs are largely independent of the substrate chosen.
4. Finally, a ring structure exhibits maximal length for a given n and k over all regular, periodic graphs.

It seems apparent then that rings are of special interest, and this suggests the form of a graph model that interpolates between similar (but not identical) limits to the α-model, but with no sociological apparatus required to motivate it. In this model—the β-model—there is no talk of mutual friends or clusters or acquaintance circles, but simply a perfect ring structure that, by virtue of a single parameter, metamorphoses into a random graph.

The Model

The algorithm for the β-model starts with a perfect 1-lattice, in which each vertex has precisely k neighbours ($k/2$ on either side), and then *randomly rewires* the edges of the lattice, with probablility β, using the following algorithm:

1. Each vertex i is chosen in turn, along with the edge that connects it to its nearest neighbour in a clockwise sense $(i, i+1)$.
2. A uniform random deviate r is generated. If $r \geq \beta$, then the edge $(i, i+1)$ is unaltered. If $r < \beta$, then $(i, i+1)$ is deleted and *rewired* such that i is connected to another vertex j, which is chosen uniformly at random from the entire graph (excluding self-connections and repeated connections).
3. When all vertices have been considered once, the procedure is repeated for edges that connect each vertex to its *next-nearest neighbour* (that is, $i+2$), and so on. In total $k/2$ such *rounds* are completed, until all edges in the graph have been considered for rewiring exactly once.

Hence when $\beta = 0$, the resulting graph remains precisely a 1-lattice, and when $\beta = 1$, all edges are rewired randomly, resulting in a close approximation to a random graph.[7] The intermediate cases ($0 < \beta < 1$) are less easily understood, but their interpretation seems much clearer than the equivalent graphs constructed using the α-model. That is, the β-algorithm starts with a rigidly ordered graph with completely known properties and then gradually transforms it into what is effectively a random graph, the properties of which are known asymptotically. Thus the β-model interpolates between more definite endpoints than the α-model, and the action of the parameter can be interpreted more easily as a variable degree of stochasticity in the graph-construction algorithm (see Fig. 3.19). There is no sociological justification to this model, which is precisely the point, the question being Do the same transitions in the statistical properties of the α-graphs occur in β-graphs? If so, what does this say about the underlying cause and generality of the α-graph properties?

Length and Clustering Properties

The length and clustering measurements of Section 3.1.1 can now be repeated for β-graphs, over the parameter range $0 \leq \beta \leq 1$ and averaged

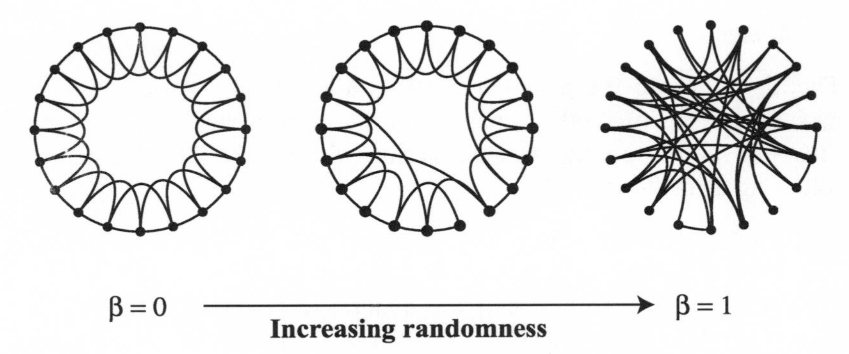

Figure 3.19 Schematic of the β-graph construction algorithm. For $\beta = 0$, the original 1-lattice is unchanged; for $\beta = 1$, all edges are rewired randomly; and for $0 < \beta < 1$, graphs combining elements of order and randomness are generated.

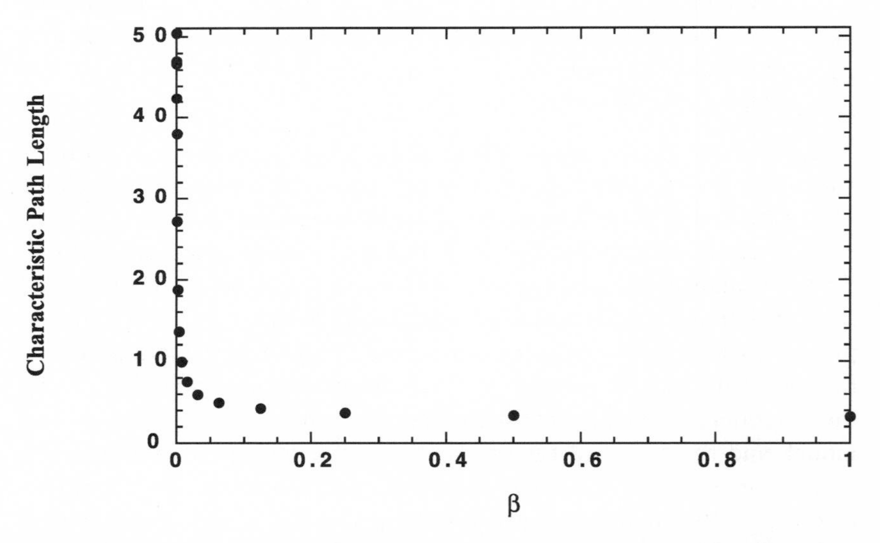

Figure 3.20 L vs. β for β-graphs averaged over twenty realisations of the construction algorithm ($n = 1{,}000$, $k = 10$).

over twenty realisations of the construction algorithm. As was the case with the α-model, a sudden and rapid transition occurs in $L(\beta)$ between the ring and random graph extremes (Fig. 3.20). However, all the action occurs very close the to $\beta = 0$ limit, and so, in order to resolve the detailed behaviour, it is necessary to plot $L(\beta)$ on a semilog scale (Fig. 3.21). The qualitative features of $L(\beta)$ are somewhat different from

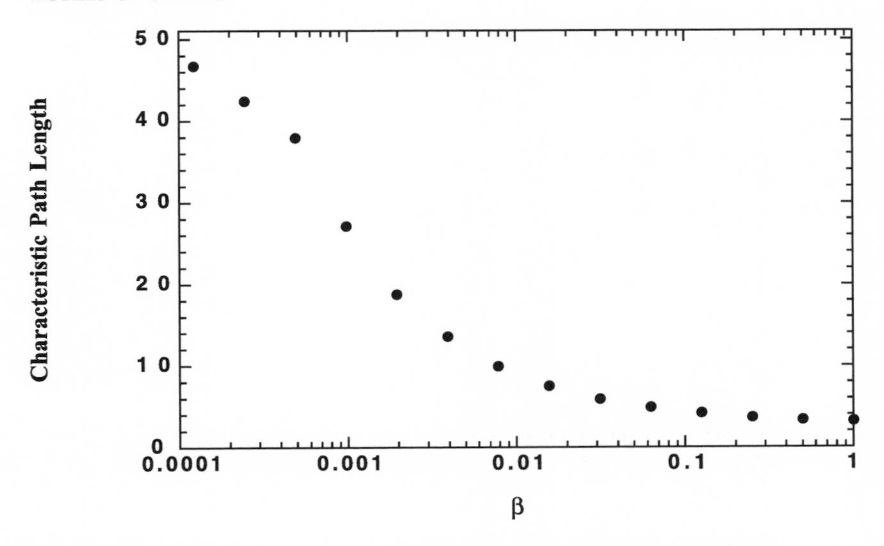

Figure 3.21 Averaged L vs. β viewed on a semilog scale for β-graphs ($n = 1{,}000$, $k = 10$).

$L(\alpha)$, yet the "cliff" in the small-β region appears to correspond to the cliff in the α-model, and the length scaling in this region also appears to be logarithmic with respect to n. The relationship between $L(\beta)$ and $\gamma(\beta)$ is also similar, the main point being that $\gamma(\beta)$ remains high long after $L(\beta)$ has approached its random graph asymptote (Fig. 3.22). Obviously, however, the transitions in L and γ for the two models occur at quite different values of the respective model parameters.

How to Compare?

This mix of similarity and dissimilarity between the two models raises the issue of how they can be compared. Clearly both models exhibit similar limiting properties, and both also display sudden and rapid transitions between these limits. But it is not clear how the two model-specific parameters (α and β) relate to each other, if at all. In other words, are the observed properties of the two models generated by different underlying mechanisms? Or are they simply different manifestations of the same mechanism, which can be revealed by identifying the appropriate, model-independent parameter? In the next section we shall see that the latter is in fact the case and that, as a result, the properties required to generate small-world features in a graph can be abstracted to avoid even the limited sociological descriptors utilised in the α-model.

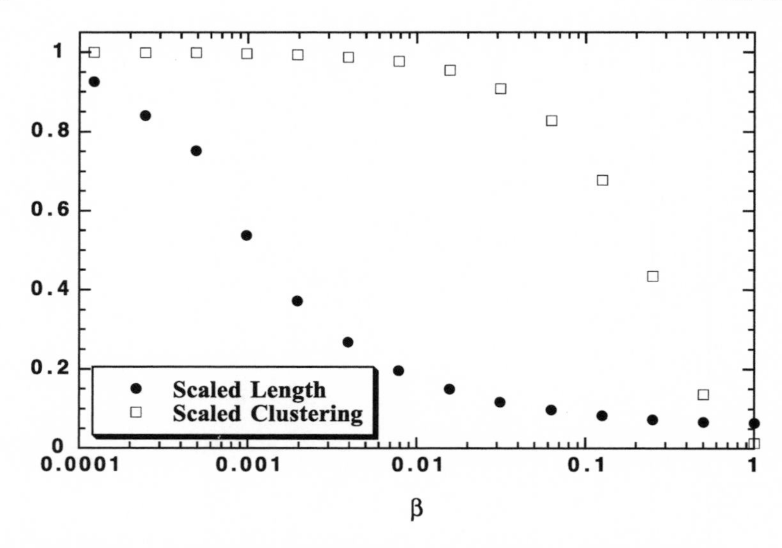

Figure 3.22 Scaled $\gamma(\beta)$ and $L(\beta)$ for β-graphs averaged over twenty realisations ($n = 1{,}000$, $k = 10$).

3.1.3 Shortcuts and Contractions: Model Invariance

In order to understand the mechanism controlling the length properties of α- and β-graphs, notice that in both cases a low parameter value corresponds to a low degree of randomness, in that new edges are highly correlated either with existing edges or with a predefined 1-lattice structure. In either case, edges are disproportionately likely to connect vertices that already share at least one mutually adjacent vertex. That is, at low parameter values, new edges (i, j) tend to form *triads* (see Fig. 3.23). These new edges therefore supersede, as the shortest path between i and j, a path with a length that was only two in the first place. In contrast, at high parameter values, edges are just as likely to form between vertices that have nothing in common, may come from opposite ends of the 1-lattice, or are uncorrelated with any existing edges. Such edges tend not to form triads and so create direct links between vertices previously separated by path lengths greater than two (see Fig. 3.23). This observation motivates the following definitions.

Definition 3.1.1. The *range* of an edge $R(i, j)$ is the length of the shortest path between i and j in the *absence* of that edge.

In other words $R(i, j)$ equals the *second*-shortest path length between i and j.

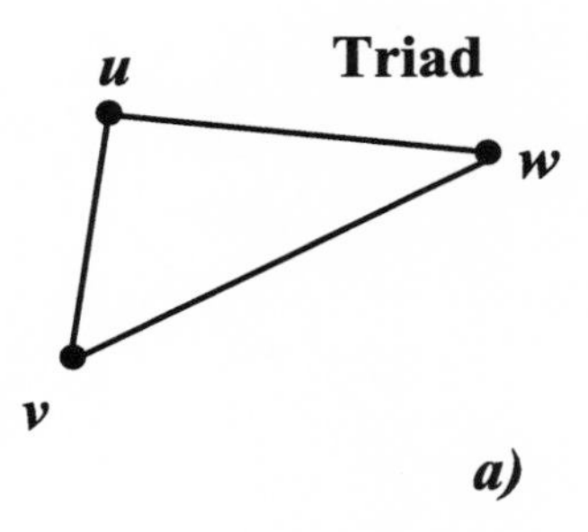

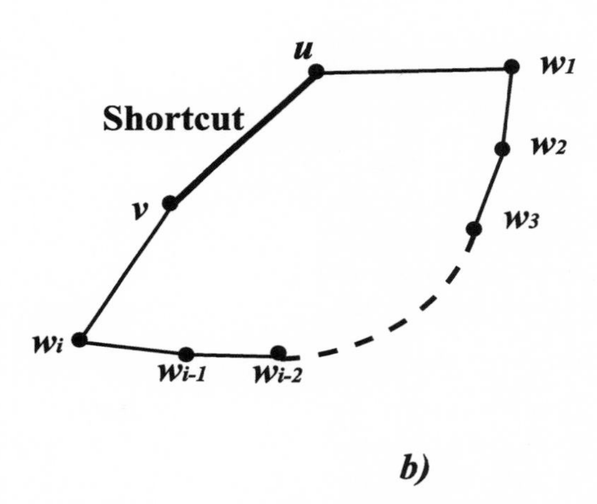

Figure 3.23 Most edges in a highly clustered graph form triads (a), but some connect vertices that would otherwise be widely separated (b). These latter edges are called *shortcuts.*

Definition 3.1.2. An edge (i, j) is called an *r-edge* if it has a range $R(i, j) = r$.

According to this definition, edges that complete triads are 2-edges, and edges that do not complete triads are r-edges with $r > 2$. Because these edges will turn out to be significant, they merit a special name that implies something of their function.

Definition 3.1.3. An r-edge with $r > 2$ is called a *shortcut.*

For both α- and β-graphs, $r > 2$ is the natural threshold for an edge to be classified as a shortcut: edges that connect vertices within the same neighbourhood must have $r = 2$, and edges connecting vertices not in

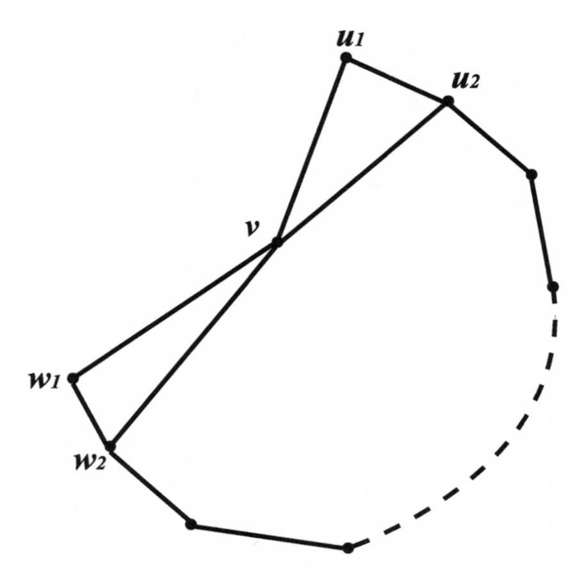

Figure 3.24 The simplest example in which the path length between pairs of vertices can be contracted by a combination of edges, none of which are shortcuts. The vertex responsible is called a *contractor*.

the same neighbourhood are just as likely to connect vertices anywhere in the graph. Hence r is either two or else a uniform random variable whose mean is on the order of L. Notice also that, although the two models motivated the idea of the range of an edge, its definition does not require any information about when, in the construction process, the edge was built or what model was used to build it. Hence the following *model-independent* parameter can be defined:

Definition 3.1.4. Given a graph of $M = (k \cdot n)/2$ edges, the fraction of those edges that are shortcuts is denoted by ϕ.

It turns out that shortcuts are instrumental in contracting the characteristic path length of a graph, but, before even seeing the evidence for this, it is already possible to see a potential problem. There is at least one circumstance under which a graph can have its length contracted without any shortcuts present, the simplest example of which is shown in Figure 3.24. The vertices u_i and w_i clearly would be widely separated were it not for the edges adjacent with v, yet none of these edges is a shortcut according to the definition. This situation, far from being improbable, is one we might well expect to encounter in real social networks in which otherwise disparate *groups* are connected by a single party who is a member of each. In fact, a shortcut is simply the most extreme case of this situation, where at least one of the groups consists of only a single member. It seems clear then that length contraction should be possible without shortcuts, but the appropriate definition is hard to pin down, at least in terms of edges, because in such a case, length contraction results not from any single edge, but from the combined effect of edges all emanating from the vertex v. Hence the natural definition is expressed in terms of vertices that share not an edge, but a neighbourhood.

Definition 3.1.5. If two vertices u and w are both elements of the same neighbourhood $\Gamma(v)$, and the shortest path length between them that does not involve any edges adjacent with v is denoted $d_v(u, w) > 2$, then v is said to *contract* u and w, and the pair (u, w) is said to be a *contraction*.

Equivalently a contraction results whenever a pair of vertices that *are not themselves connected* have one and only one common neighbour.

Definition 3.1.6. ψ is the fraction of all pairs of vertices that are not connected and have one and *only* one common neighbour.

Hence ψ is an analogous parameter to ϕ, although it is more general, as most shortcuts result in contractions but not the reverse. Nevertheless, both parameters quantify a similar notion—the contraction of distance between previously widely separated parts of a graph.

Comparison of α- and β-Models Using Shortcuts

It is now possible to recast the two models (α-graphs with a ring substrate and β-graphs) in terms of the new parameter ϕ. It is clear that for very small ϕ, the two models produce different values of $L(\phi)$ and $\gamma(\phi)$ (see Figs 3.25 and 3.26). This discrepancy is not surprising given the obvious differences between the two models. What is more surprising is that, as ϕ increases, the statistics of the models rapidly converge and are virtually identical well before the $\phi \approx 1$ random limit is reached. This coincidence can be understood by considering the functional dependence of ϕ upon the two model parameters α and β. In the α-model, $\phi(\alpha)$ remains virtually constant for small α and then rises suddenly and dramatically to a new plateau (Fig. 3.27), whereas $\phi(\beta)$ rises more or less linearly from 0 to almost 1 (Fig. 3.28). That $\phi < 1$ in both cases, for all parameter values, is due to the fact that, even in a random graph,

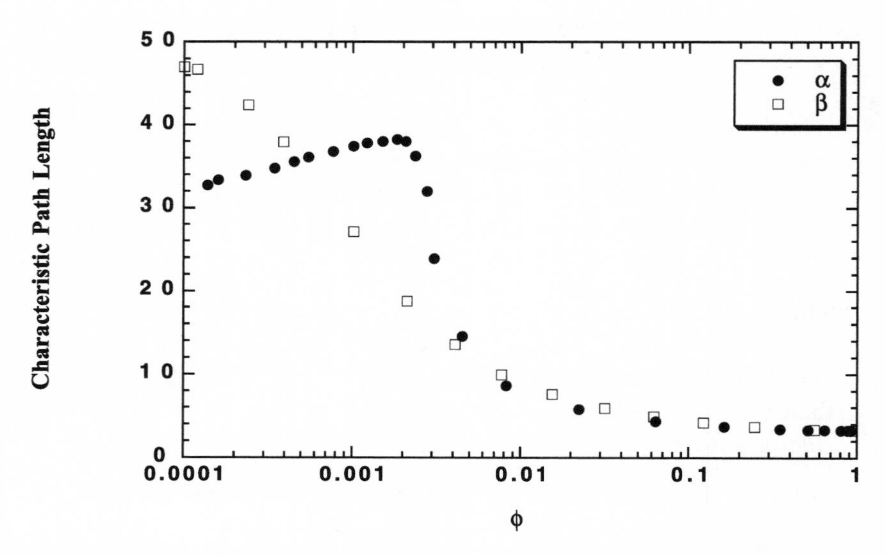

Figure 3.25 Comparison of L vs. fraction of shortcuts ϕ for α- and β-graphs ($n = 1{,}000$, $k = 10$).

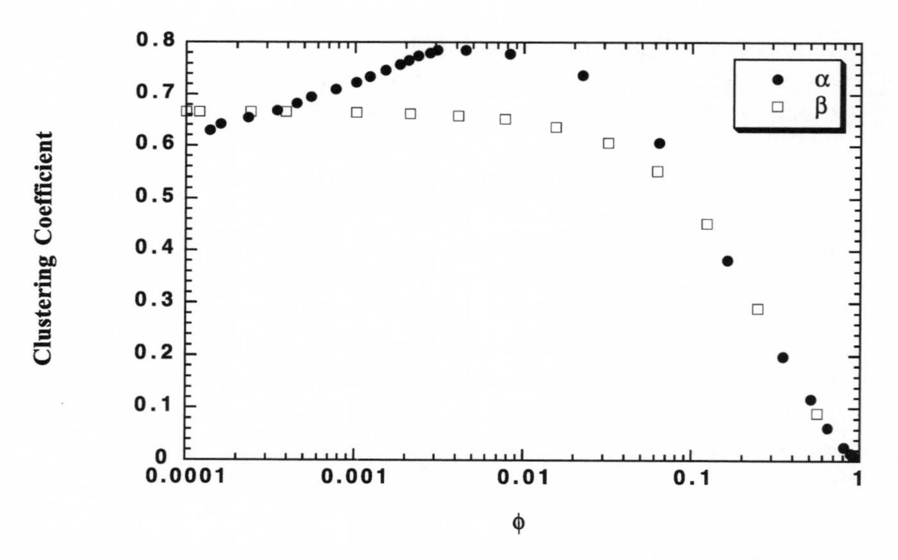

Figure 3.26 Comparison of γ vs. fraction of shortcuts ϕ for α- and β-graphs ($n = 1{,}000$, $k = 10$).

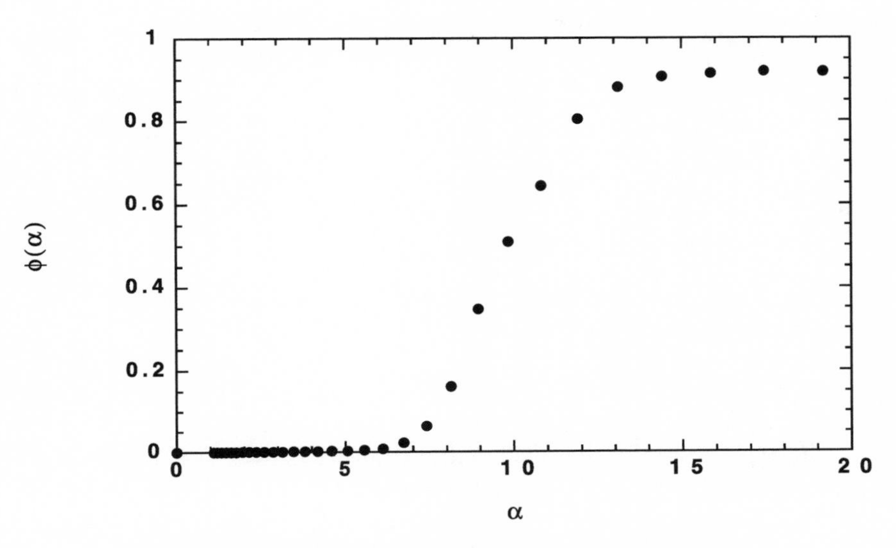

Figure 3.27 $\phi(\alpha)$ vs. α for α-graphs ($n = 1{,}000$, $k = 10$).

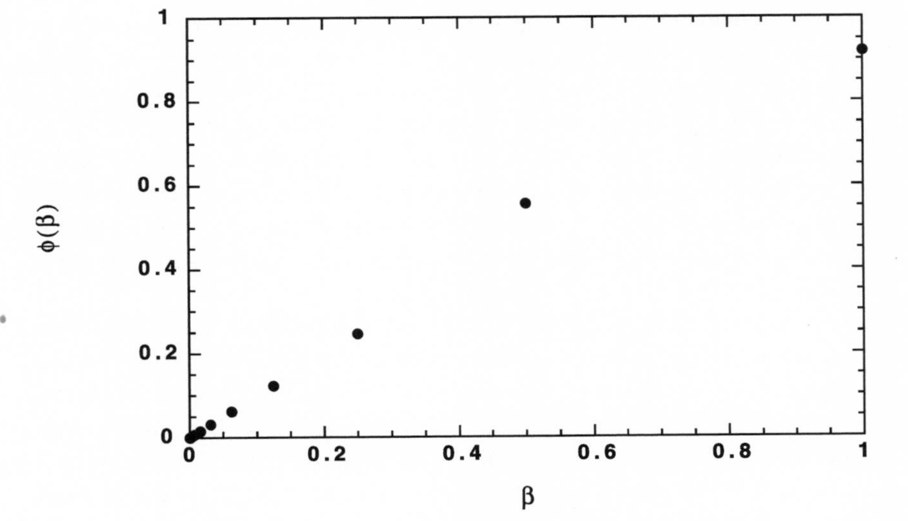

Figure 3.28 $\phi(\beta)$ vs. β for β-graphs ($n = 1{,}000$, $k = 10$).

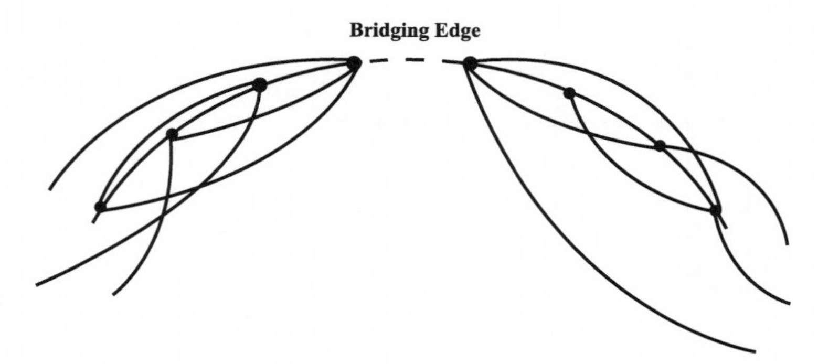

Figure 3.29 A possible example of a bridging edge, which occur in α-graphs, especially for low k.

there is still some residual probability that triads will be created occasionally.

The explanation for the small-ϕ discrepancies between the two models is illustrated in Fig. 3.29: additional edges added on to a ring substrate in the α-model can form exclusively to the right and left of one of the substrate edges, causing it to be classified as a shortcut, despite the fact that it does not connect distant vertices. These anomalies (known as *bridging edges*) become less significant as k increases, and it becomes increasingly likely that at least one edge will cross the shortcut, reducing it to become a 2-edge. This irregularity is inherent to the α-model and is one reason why the β-model is easier to deal with: such anomalies cannot occur for low β, and so the only shortcuts registered are "genuine" in the sense that every new shortcut connects previously distant parts of the graph (hence the strict monotonicity of the $L(\beta)$ curve, parameterised by ϕ). The discrepancy between the $\gamma(\phi)$ values persists for values of ϕ that are significantly greater than in the $L(\phi)$ comparison—a reflection of the local nature of the clustering coefficient.

At this point, it is worth emphasising what is going on here. These two types of graphs have little superficially in common, except perhaps at the extremes of their parameter ranges. The α-graphs are considerably more irregular than their low-β counterparts, and, even though it must be true that both converge to approximations of the random graph, nowhere is it obvious that this has to be governed by a statistic like ϕ, or that it must be governed in a similar fashion in each model. In fact, ϕ is a crude statistic that appears to miss much of what might cause the characteristic path length of a graph to shrink. Indeed for sufficiently small ϕ, the models cannot be reconciled. Nevertheless, it appears to capture a remarkable amount of the structural changes that occur in the transition from a highly clustered, one-dimensional structure to that of a random graph.

The other major point to note is that almost all of the changes that occur in $L(\phi)$ do so for $\phi \ll 1$, whereas the same cannot be said for $\gamma(\phi)$. This, in fact, is crucial because it reinforces the earlier claim about the existence of small-world graphs. Even a relatively small fraction of shortcuts (say, $\phi \approx 0.01$ for $n = 1{,}000$, $k = 10$) is sufficient to bring the graph to its asymptotic value of L, yet it remains highly clustered and, from a local perspective, much more like its original structure than that of a random graph. A priori, this was not obvious, yet the reason is clear on reflection: when only a few shortcuts exist in a large graph, every shortcut (created at random) is likely to connect widely separated parts of the graph and thus, on its own, have a significant impact upon the characteristic path length of the entire graph. Each shortcut contracts not just the distance between the two vertices in question, but between their two neighbourhoods, their neighbourhoods' neighbourhoods, and so on, generating a highly nonlinear effect. As more and more shortcuts are created, they must necessarily have less and less impact, because the graph itself has become effectively smaller, hence the asymptotic nature of $L(\phi)$. Meanwhile, the removal of an edge from a triad has relatively little effect on $\gamma(\phi)$, it being just one edge amongst many. The result is a large interval of graphs with small L and large γ: small-world graphs.

Substrates Revisited

So far the use of a ring substrate has been justified from a number of perspectives, but the question remains to what degree any graph model based on an apparently special choice of substrate can be held to be generically representative of a truly broad class of graphs, ranging from ordered to random. Can the notion that shortcuts largely determine the characteristic path length of a graph help to unify the results obtained for the ring substrate of the α-model with its alternatives?

It turns out (of course) that it can. Figure 3.30 shows a comparison of $L(\phi(\alpha))$ generated for $n = 1{,}000$, $k = 10$ for each one of the substrates discussed earlier, with the exception of the two-dimensional-lattice substrate, which (because it must be a perfect square) has $n = 32^2 = 1{,}024$. Three features of this picture are particularly noteworthy:

1. Although the various $L(\phi(\alpha = 0))$ for different substrates span over an order of magnitude (and, as we have seen, exhibit different

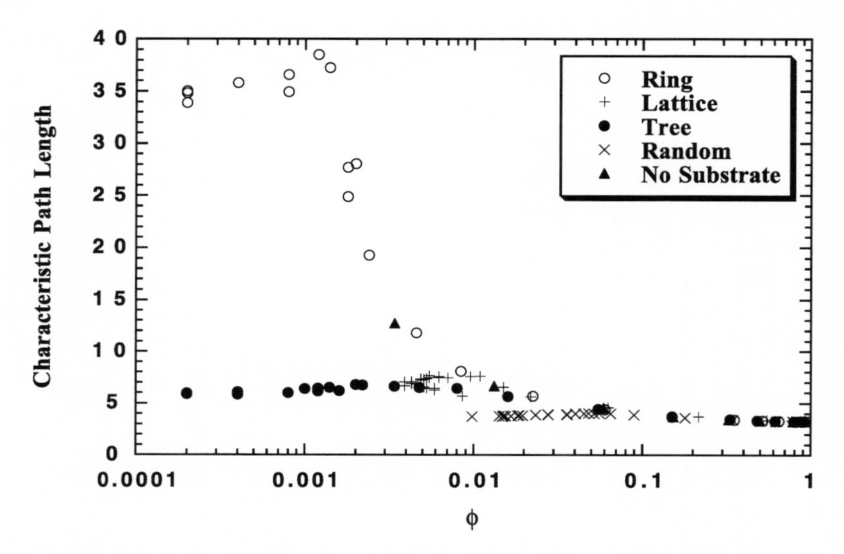

Figure 3.30 Comparison of L vs. ϕ for all substrates of the α-model, with $n = 1{,}000$, $k = 10$ (except for two-dimensional lattice, which has $n = 1{,}024$).

length-scaling laws at $\phi = 0$), all such curves converge to the same limit for small ϕ (for the parameters used, the value is $\phi \approx 0.01$). This is an important point: as long as ϕ is not too small, then it is irrelevant what sort of substrate the model might have been constructed from, because its length (but not clustering) properties are essentially that of a random graph.

2. In the case of the random substrate, $\phi(\alpha) > 0$ for all α. In other words, once a random graph becomes connected (at $k \approx \ln(n)$), then it is already "small," and the subsequent addition of correlated edges upon such a substrate does little to negate its random nature.
3. Once the no-substrate α-model becomes connected, its length is indistinguishable from that of the ring-substrate α-model. This supports the claim that when the disconnected "caveman" world becomes connected, it does so according to the topology of a ring.

A parallel situation exists for $\gamma(\phi)$, except that here (Fig. 3.31) the correspondence between the different substrates is not as close, with the agreement especially poor for the random substrate. Nevertheless, all substrates clearly admit small-world graphs, and, for $\phi \gtrsim 0.1$, all differences due to the topology of the substrates have been overwhelmed *but still before* the random limit has been reached.

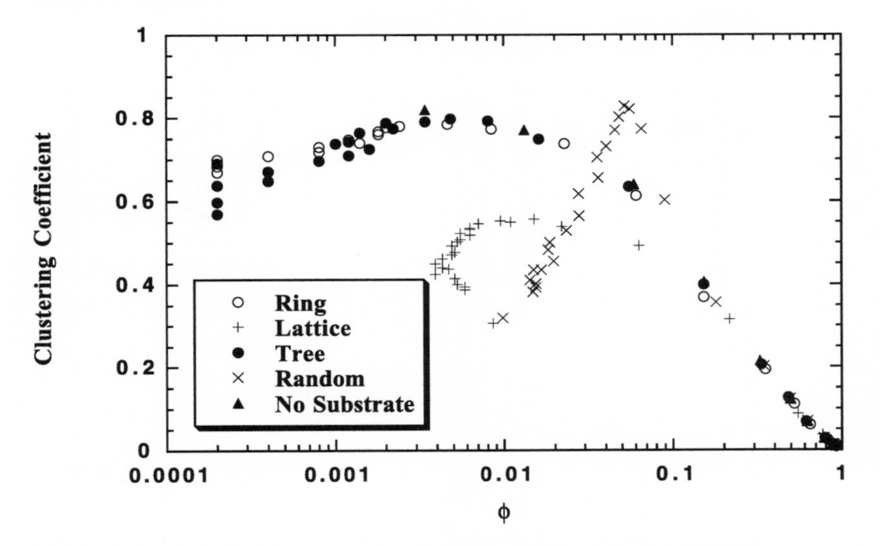

Figure 3.31 Comparison of γ vs. ϕ for all substrates of the α-model, with $n = 1{,}000$, $k = 10$ (except for two-dimensional lattice, which has $n = 1{,}024$).

On the basis of these observations it is at least plausible to claim that if our network of interest: (1) is connected, and (2) has a ϕ that is not too small,[8] then we need not be concerned with the issue of the substrate because *all* substrates yield essentially the same result—that the network can be highly clustered and still be small.

Comparison of α- and β-Models Using Contractions

The same comparisons between α- and β-graphs and between different substrates of the α-model can also be made with respect to ψ. Figures 3.32–3.35 show mixed results.[9] On the one hand, the different α-substrates show closer agreement with respect to ψ than ϕ (Figs. 3.32 and 3.33), but, on the other hand, much poorer agreement is attained for the corresponding comparison of the α-model with a ring substrate and the β-model (Figs. 3.34 and 3.35). The explanation for both these phenomena is related to the fact that all graphs considered have nonzero ψ for all values of their respective parameters. For instance, even in a perfect 1-lattice (e.g., for $\beta = 0$) one pair of vertices in every neighbourhood must necessarily be contracted (the two vertices at the outer limits of the neighbourhood), leading to a nonzero $\psi = 2/k(k-1)$. These "substrate" contractions clearly do little to influence the characteristic

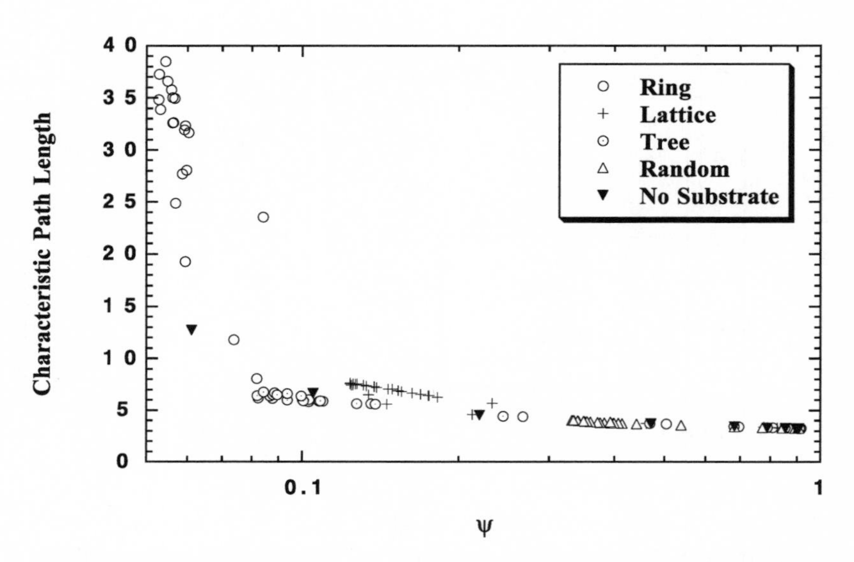

Figure 3.32 Comparison of L vs. ψ for all substrates of the α-model, with $n = 1{,}000$, $k = 10$ (except for two-dimensional lattice, which has $n = 1{,}024$).

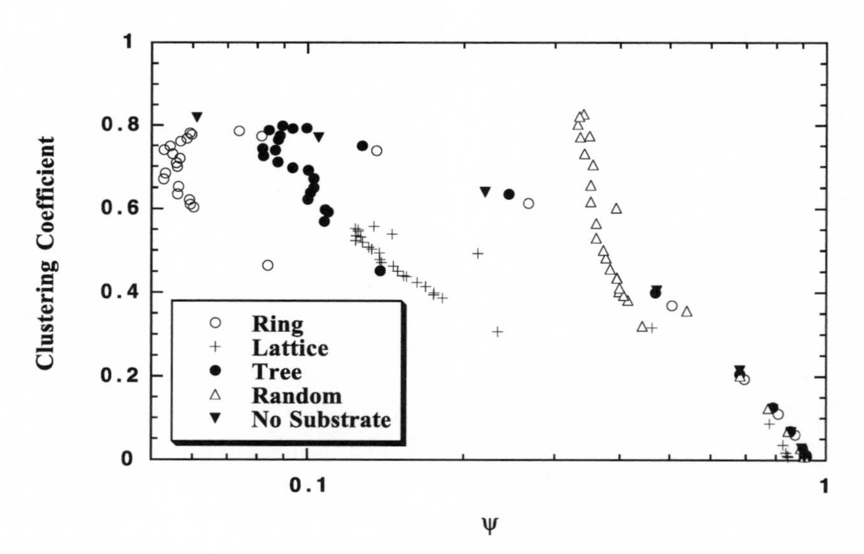

Figure 3.33 Comparison of γ vs. ψ for all substrates of the α-model, with $n = 1{,}000$, $k = 10$ (except for two-dimensional lattice, which has $n = 1{,}024$).

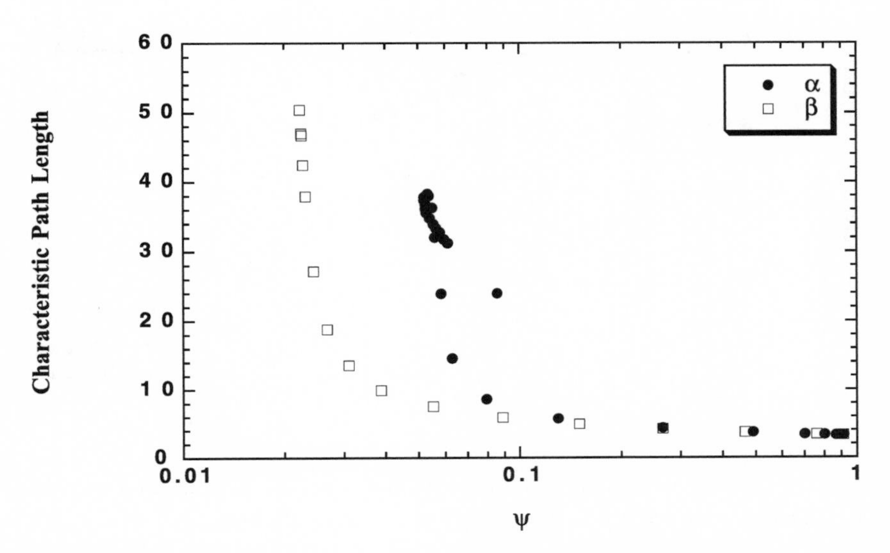

Figure 3.34 Comparison of L vs. ψ for α- and β-graphs ($n = 1{,}000$, $k = 10$).

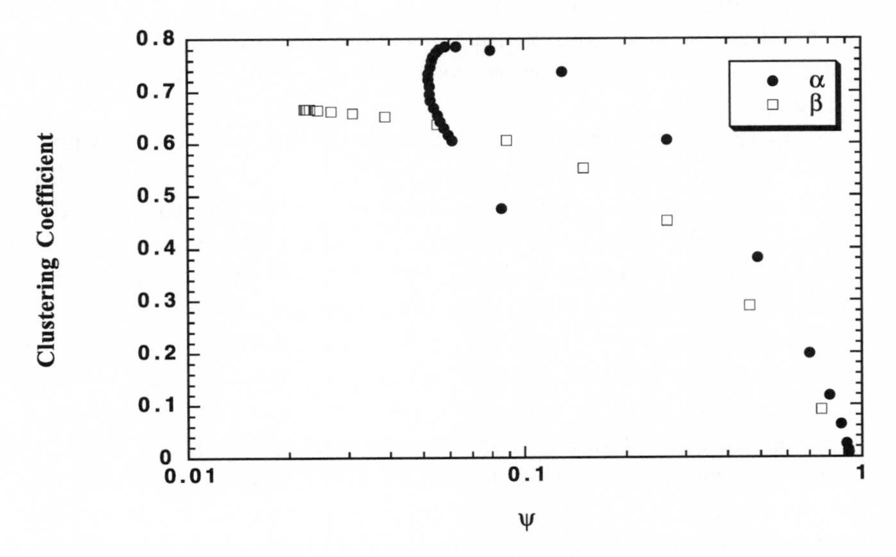

Figure 3.35 Comparison of γ vs. ψ for α- and β-graphs ($n = 1{,}000$, $k = 10$).

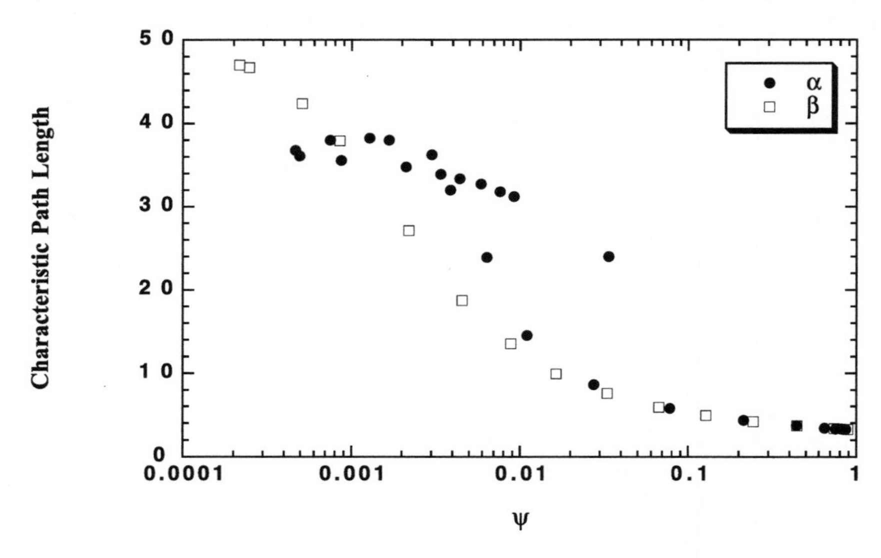

Figure 3.36 Comparison of L vs. ψ for α- and β-graphs, with substrate contractions subtracted. The comparison improves signficantly.

path length of a β-graph, and one might be tempted to factor them out, thus aligning the α- and β-models (Fig. 3.36), except that it appears to be precisely the analogous contractions that result in the observed length differences between the various α-substrates and that should therefore be retained.

At this point, then, it seems that no single statistic can adequately unify all the types of graphs that have been considered. Furthermore, the situation will become even more complicated and ambiguous with the inclusion of spatial graphs under the same theoretical umbrella. Hence it is difficult to say much more than that both shortcuts and contractions appear to have significant bearings on the length and clustering properties of relational graphs, and that both elucidate features that are common to all graphs in this category, without capturing all of the structural nuances associated with different construction algorithms and substrates. Perhaps it is too much to ask of *any* single statistic that it perform this feat: certainly the existence of such a statistic appears to be an open question. In the absence of a better alternative then, shortcuts and contractions will have to suffice. Because both perform the function of bringing otherwise distant parts of the graphs close together, given the conceptually simpler nature of shortcuts, these will be used exclusively, except where they cannot provide sufficient explanatory power.[10]

Length Scaling Revisited: Motivation for ϕ-Model

As mentioned earlier, it is not the specific value of a graph's characteristic path length that is interesting, so much as its length relative to that of a random graph with the same n, k. Along with this comparison goes the length-*scaling* properties of the graph. The reason for this is that length scaling is a topological property and so characterises a type of graph rather than any one particular graph. The problem that arises in measuring the length-scaling properties of either α- or β-graphs as a function of ϕ is that, because ϕ is itself a function of the actual, model-specific parameter, it is impossible to ensure a fixed ϕ for graphs of different n and k and hence determine their length-scaling behaviour. What is required is a graph model that preserves the features of the α- and β-models but for which ϕ can be specified explicitly rather than parametrically. This can be achieved by the following construction:

1. Construct a 1-lattice with n vertices in which every vertex has degree k.
2. Specify a desired ϕ.
3. "Randomly rewire" the 1-lattice in much the same way as was done for the β-model, but with the additional constraint that $\phi \cdot (k \cdot n)/2$ of its edges be forced to be shortcuts.[11] Specifically:
    a. Pick a vertex u at random.
    b. Pick immediate neighbours v at random until one is found such that u and v have a mutual friend (in other words, the edge (u, v) is not already a shortcut).
    c. Delete the edge (u, v).
    d. Choose vertices w at random with uniform probability over the entire graph, until one is found that does *not* have any mutual friends with u (that is, (u, w) *is* a shortcut).
    e. Add the edge (u, w).

In this manner, *ϕ-graphs* can be constructed with any desired fraction of shortcuts for any n and k. Hence the scaling properties of graphs with fixed ϕ can be studied.[12]

Properties of the ϕ-Model

Before proceeding to analyse the ϕ-model, it is first necessary to verify that it does indeed produce results akin to those generated by α- and β-graphs for the same parameter values. Fig. 3.37 shows the comparison

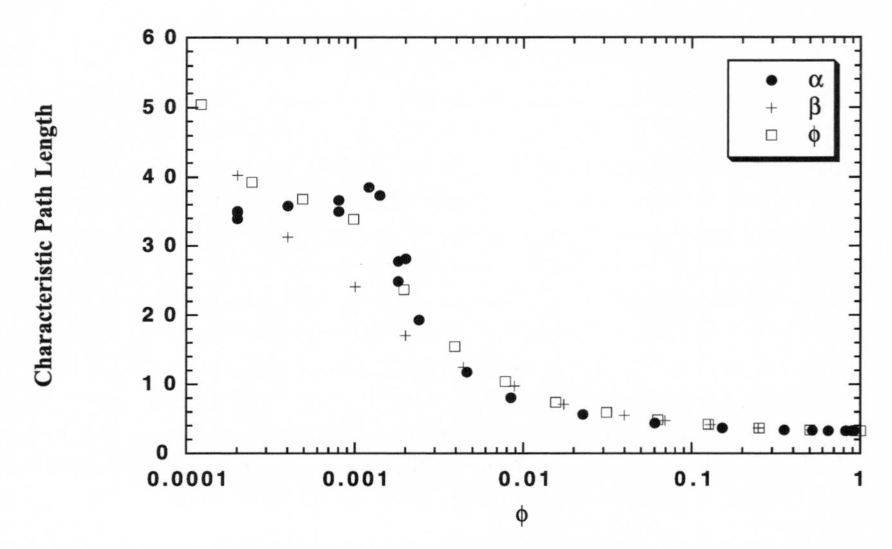

Figure 3.37 Comparison of L vs. ϕ for α-, β-, and ϕ-graphs ($n = 1{,}000$, $k = 10$).

for characteristic path length and reveals that ϕ-graphs appear to share features with both previous models but are more similar to β-graphs, as one would expect given the similarity of the respective algorithms. Reasonable agreement also exists for the other statistic of interest: $\gamma(\phi)$ (Fig. 3.38). It seems then that the ϕ-model is an acceptable device for studying the properties of small-world graphs.

Figures 3.39, 3.40, and 3.41 show $L(\phi)$ for ϕ-graphs at increasing values of n, for a fixed $k = 10$. The important point to notice, besides the usual cliff structure converging to a random-graph limit, is that at low n and sufficiently small ϕ, there is a plateau prior to the cliff. The reason for this is straightforward: the cliff commences only when the first shortcut is created. Hence, for a given n and k, if $\phi \geq 2/k \cdot n$ then at least one shortcut will exist in the ϕ-graph, and $L(\phi)$ will be strictly less than $L(\phi = 0)$, which is the 1-lattice extreme. Obviously, however small ϕ is made, for sufficiently large n at least one shortcut will exist. Fixing $\phi = \phi_* > 0$ and then considering only graphs with $n > n_{\min} = \phi_* \cdot k/2$, Figure 3.42 shows that the relevant graphs scale logarithmically. More importantly, this appears to be true for arbitrarily small ϕ_*, as long as $n_{\min}$ is set sufficiently high. These observations lead to the following conjecture.

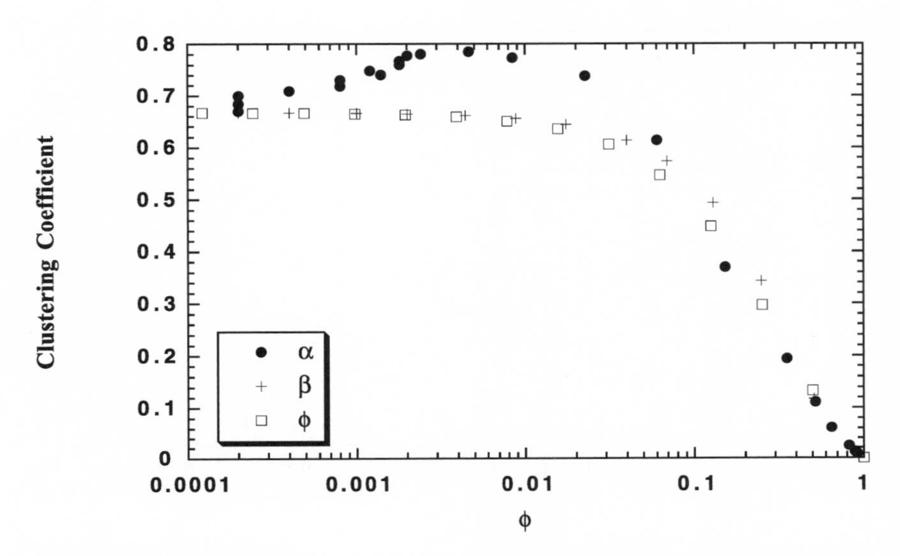

Figure 3.38 Comparison of γ vs. ϕ for α-, β-, and ϕ-graphs ($n = 1{,}000$, $k = 10$).

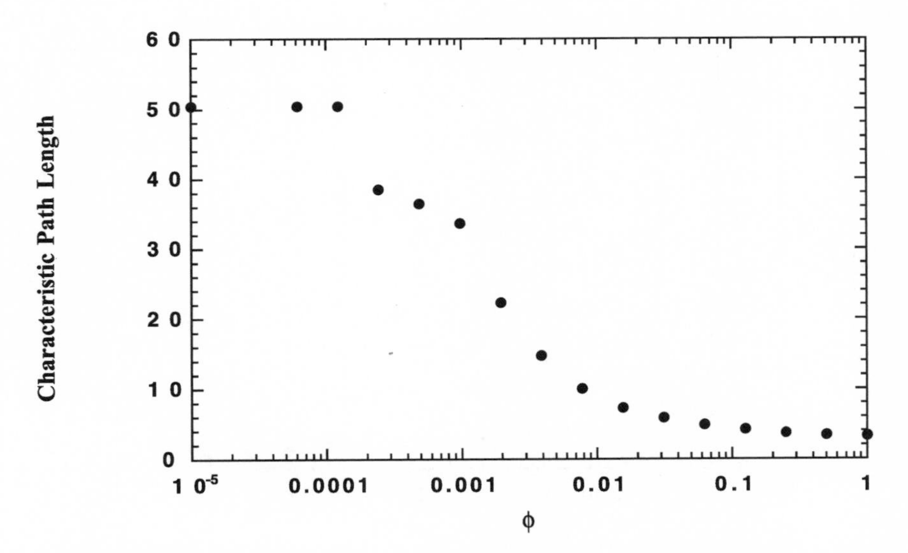

Figure 3.39 L vs. ϕ for ϕ-graphs ($n = 1{,}000$, $k = 10$). A high plateau is clearly visible before cliff.

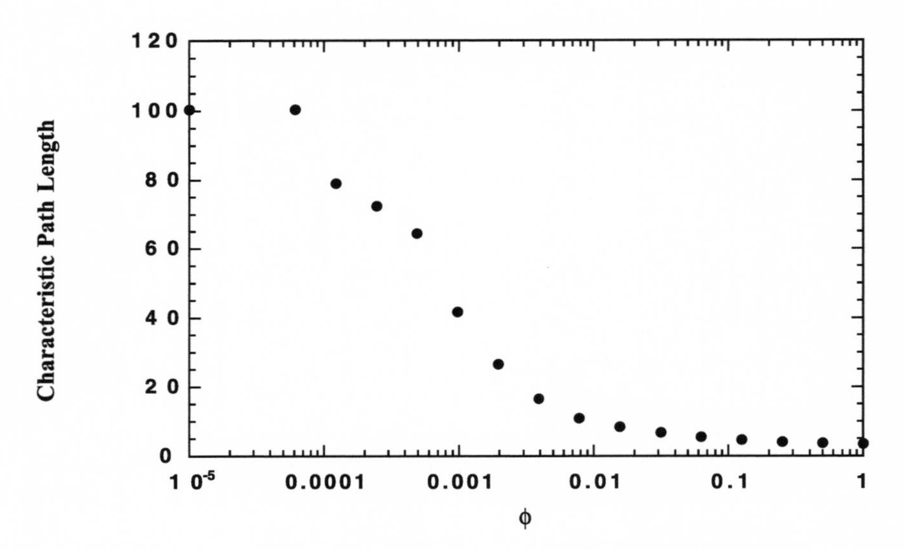

Figure 3.40 L vs. ϕ for ϕ-graphs ($n = 2{,}000$, $k = 10$). Plateau is still visible before cliff but ends for lower ϕ.

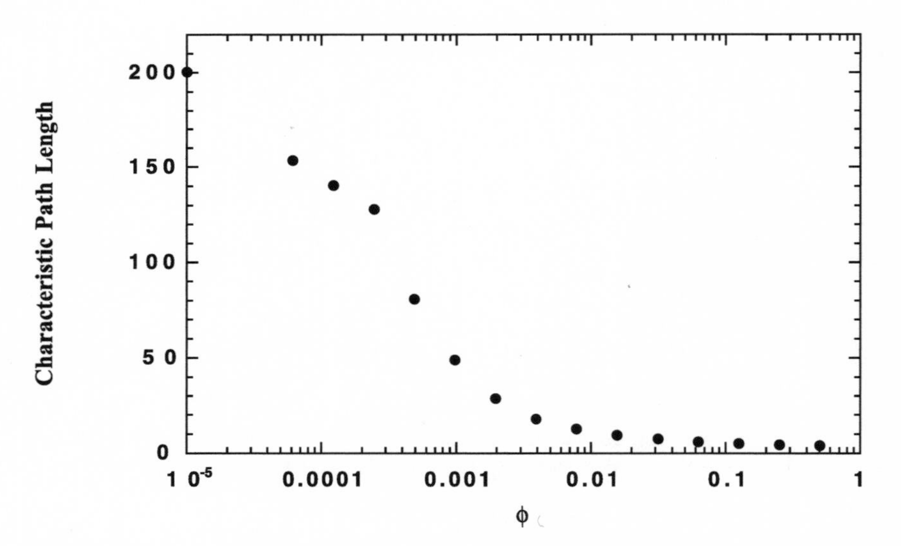

Figure 3.41 L vs. ϕ for ϕ-graphs ($n = 4{,}000$, $k = 10$). Plateau is no longer visible.

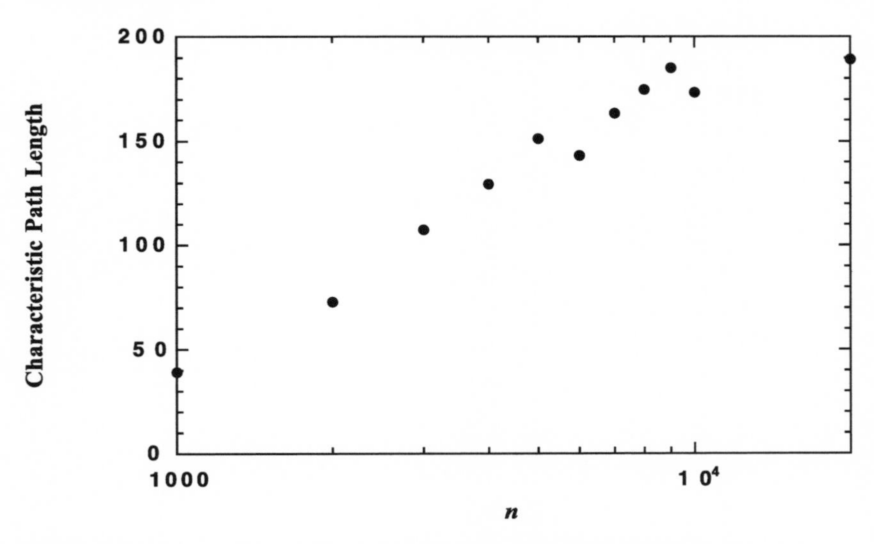

Figure 3.42 L vs. n for ϕ-graph with $k = 10$, $\phi \approx 0.00024$ (the value of ϕ at which the plateau ends for $n = 1{,}000$).

Conjecture. *For any $\phi_* > 0$ and for $n > 2/(k\phi_*)$, ϕ-graphs with fixed $\phi = \phi_*$ will exhibit logarithmic length scaling with respect to n (where $n \gg k \gg 1$).*

How this occurs is the subject of Chapter 4, but basically the idea seems to be that once n_{min}, ϕ_*, and k are picked such that $\phi_*(n_{\text{min}} \cdot k)/2 \geq 1$, then, as n increases, more and more shortcuts appear, causing the length to grow more slowly than it otherwise would for a one-dimensional structure. That it should grow logarithmically is not obvious; much of Chapter 4 is devoted to understanding this.

3.1.4 Lies, Damned Lies, and (More) Statistics

Although the idea of shortcuts in a graph appears to capture some interesting properties, it is still a crude statistic. In fact, there are two natural ways in which the notion of a shortcut can be generalised to capture, in a more detailed fashion, the transition from ordered to random graphs. Although neither of these more general notions seems to reveal any new features in the $L(\phi)$ and $\gamma(\phi)$ behaviour, they do lead to a new statistic, which can be thought of as a measure of the graph's *structural complexity.*

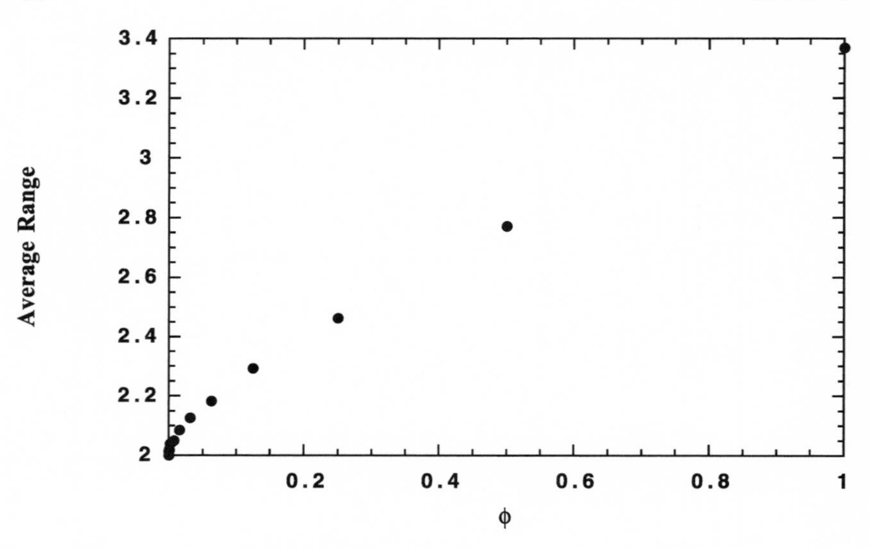

Figure 3.43 Average range R vs. ϕ for ϕ-graphs ($n = 1{,}000$, $k = 10$).

Average Range

The definition of a shortcut in Section 3.1.3 makes a distinction only between edges with a range of 2 and edges with a range greater than 2. In fact, range is a more flexible idea than this, and it is possible to measure the *average range* R of all edges over the entire edge set of the graph. A computation of this statistic for ϕ-graphs of $n = 1{,}000$, $k = 10$ is shown in Figure 3.43. Naturally, for $\phi = 0$, $R = 2$, because every edge in the graph is a 2-edge. It is not so obvious what R should be for a random graph, but it seems reasonable to think that it would be close to the characteristic path length of the graph. That is, edges in a random graph are uncorrelated with each other, and so just because two vertices are connected does not imply that their second-shortest path length should be any shorter than average. This is obviously not true in a highly ordered graph, where two vertices being connected directly *does* imply that their second-shortest path length should be shorter than for some randomly selected pair of vertices.

Average Significance

Just as the average range of a graph is analogous to ϕ, a measure of the significance of vertices can be defined that is analogous to ψ. This notion is framed by asking the question How far apart would vertices

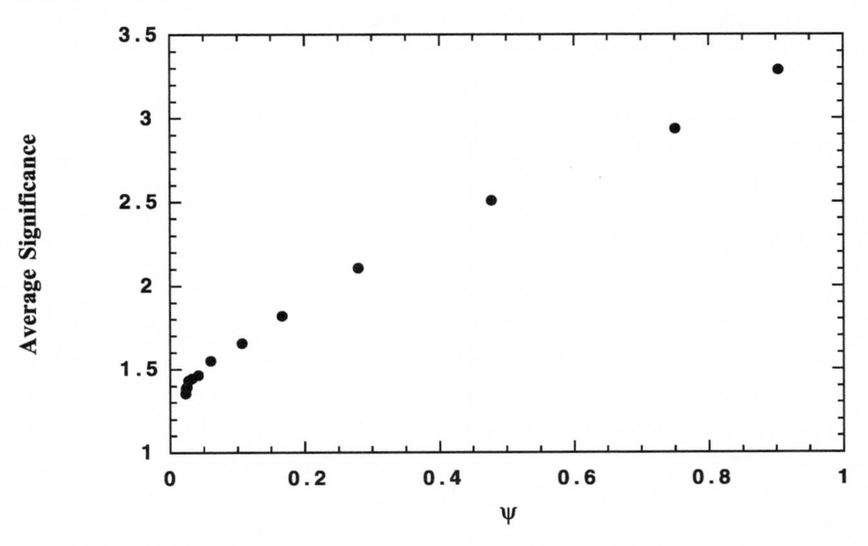

Figure 3.44 Average significance S vs. ψ for ϕ-graphs ($n = 1{,}000$, $k = 10$).

in the same neighbourhood $\Gamma(v)$ be if it were not for the connections formed by the defining vertex v? This leads to the following definition of vertex *significance*.

Definition 3.1.7. The *significance* of a vertex v is the characteristic path length of its neighbourhood $\Gamma(v)$ in the absence of v.

In other words, $S(v) = d(i, j)$ averaged over all pairs of vertices (i, j) in $\Gamma(v)$. A vertex with a significance $S(v) = s$ is called an s-vertex, and $S(v)$ averaged over all $v \in V(G)$ becomes S, the *average significance* of the graph. Hence S is the expected distance by which local neighbourhoods are separated when their central vertex is deleted. Figure 3.44 shows that the dependence of $S(\psi)$ upon ψ is qualitatively similar to that of $R(\phi)$ upon ϕ. Of particular note is that, as was the case with $R(\phi)$, S attains its maximum at the random graph limit, where it appears to approach the characteristic path length of the graph. If we think of S as a measure of a *local length scale* and L as the measure of the *global length scale*, then the convergence of these two statistics is significant in that it implies that all length scales in the graph coincide at some point at or before the random limit. In terms of how this is achieved, however, little additional insight is gained through these observations. The reason seems intuitively to be that individual shortcuts or contractions can have a highly nonlinear impact upon the characteristic path length of a graph, possibly contracting the distance between many pairs of vertices at once. In contrast, each edge or vertex is factored into the R or S calculation only once, hence (as was the case for $\gamma(\phi)$) much of the important *global* information is not picked up by these statistics.

Structural Complexity

Average range and significance are not without value, however, and may lead to a deeper understanding of the graphs in question by way of a new statistic that is not even suggested by the notion of shortcuts. This statistic which might be termed *structural complexity*, is motivated by Crutchfield's (1994) distinction between the traditional information-theoretic view of complexity, which classifies random objects as maximally complex, and what he terms *statistical complexity*, which explicitly discounts the computational effort required to generate random bits. Hence statistical complexity is related to the amount of information (think of the number of free parameters in a model) required to produce optimal forecasts of future output of a system. For a completely stochastic system, the future output can be described at best statistically, but this may require only one parameter. Hence complete randomness is, in this sense, relatively simple. Likewise, a completely periodic signal is simple because knowledge of only one parameter (the phase) is sufficient to make an optimal prediction of its future output. Crutchfield's point is that somewhere in between these two extremes, systems exhibiting a mixture of structure and unpredictability achieve a state of maximal complexity.

Structural complexity differs from Crutchfield's statistical complexity, in that it does not explicitly discount randomness. Rather, it accounts for the fact that the graphs of interest range from highly ordered to random, visiting in between a state in which an optimal description of their structure requires more information than at either extreme. This state might be expected to arise in the presence of a small fraction of long-range shortcuts (or, equivalently, highly significant vertices). Hence the following definition of *edge complexity* should capture a similar notion to statistical complexity.

Definition 3.1.8. The *edge complexity* C_e of a graph $G(\phi)$ is the average deviation of $R(\phi)$ over the entire edge set $E(G)$.

Note that *average deviation* ($s = (1/n)\sum_{i=1}^{n} |x_i - \overline{x}|$) is used instead of the more usual standard deviation as a measure of variation across a set.

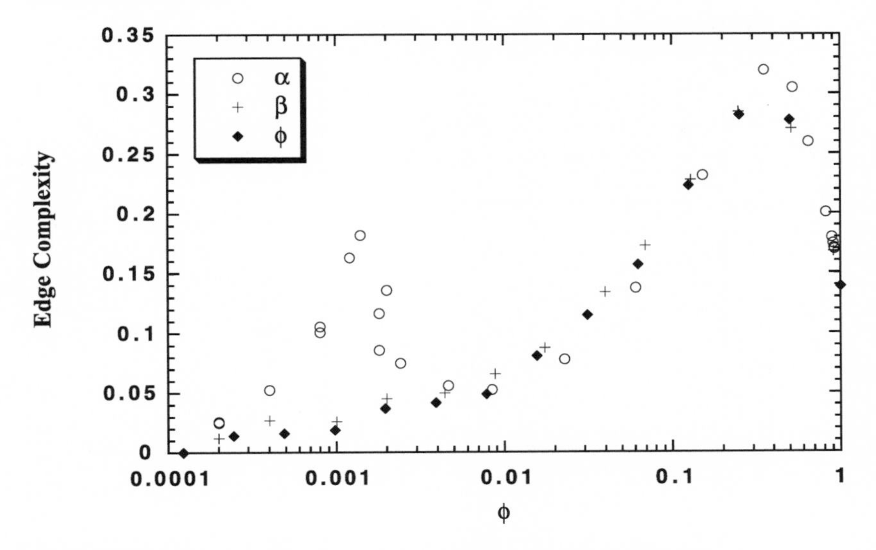

Figure 3.45 Edge complexity C_e vs. ϕ for α-, β-, and ϕ-graphs ($n = 1{,}000$, $k = 10$). The spurious hump exhibited by the α-graphs is a result of the occasional "bridging edges" that are artifacts of small k.

The reason for this is that standard deviation is especially sensitive to individual outliers in a distribution, and the nature of these graphs is that, for small ϕ, a single large-r edge can be constructed that disproportionately affects the variance. Average deviation (which takes the absolute value of deviations instead of the square) still captures large deviations from the mean but is more robust with respect to individual deviations. Figure 3.45 shows $C_e(\phi)$ for all three models considered so far. Once again, all three models show remarkable similarity (with the α-model again showing evidence of bridging edges at low ϕ), and all exhibit a distinct hump for $\phi < 1$. Furthermore, all substrates of the α-model display very similar dependencies of C_e upon ϕ (Fig. 3.46), a feature that is not at all obvious from their constructions. The interpretation of these results is still an open question.

3.2 SPATIAL GRAPHS

The second class of graph-construction algorithms to be considered follows from an entirely different premise than that of relational graphs. Instead of creating new edges as a function of preexisting edges, each vertex will be allowed to connect to other vertices according to a spec-

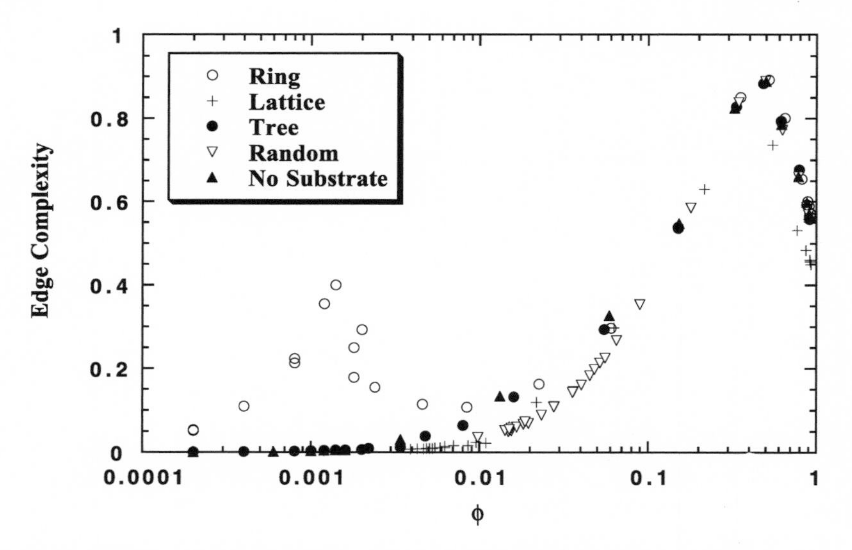

Figure 3.46 Edge complexity C_e vs. ϕ for all substrates of α-graphs ($n = 1{,}000$, $k = 10$).

ified probability distribution defined in terms of the physical distance (in $\mathbb{R}^d$) between vertices, whose locations are assumed to correspond to vertices of a cubic lattice of some specified dimension. This kind of model also generates a one-parameter family of graphs that interpolates between a d-lattice and a random graph, where the driving parameter (ξ) is some external measure of the *scale over which connections may occur*. In other words, there is presumed to be a physical force that constrains connections to within a local region. In the case of a uniform distribution in one dimension, this has a simple interpretation: $\xi = w/2$ (where w is just the width of the distribution). For a normal distribution, or any other distribution with tails, it is possible only to define an interval in which *almost all* connections will occur. For $k = O(10)$, a natural choice is $\xi = 3\sigma$ (where σ is the standard deviation of the distribution) because over 99% of all connections will be made within $\pm 3\sigma$ of the originating vertex. In the case of finite n and k, this ensures that for some finite ξ, no connections have a significant chance of occurring between vertices that are further than ξ apart. Hence the only spatial graphs to be considered here are those whose distributions exhibit (at least effectively) a *finite cutoff*.[13]

For either the uniform or Gaussian case in one dimension, an assumed uniform density of the vertices yields the condition that $\xi \geq k/2$, in order that each vertex can make k edges.[14] At the other extreme, when $\xi = O(n/2)$, edges will connect between vertices at random over the entire graph, implying that a random graph limit has been approached. Whether or not this limit is dependent upon the distribution remains to be seen, as does the nature of the transition between the ordered and random limits.

3.2.1 Uniform Spatial Graphs

The simplest kind of spatial graph is one in which vertices may connect uniformly at random to other vertices within a *spatial* distance ξ in $\mathbb{R}^d$. The detailed construction is straightforward:

1. Pick a dimension d (only $d = 1$ will be considered here).
2. For each vertex v (specified by d coordinates v_i) choose d uniform random deviates $r_i \in [-\xi, \xi]$ $(1 \leq i \leq d)$.
3. Convert the r_i to coordinates $u_i = v + \lceil r_i \rceil$.
4. Create the edge (u, v), disallowing repeated edges.
5. Go to the next vertex and repeat this procedure until $(k \cdot n)/2$ edges have been formed.

Length and Clustering

Figures 3.47 and 3.48 show L and γ versus ξ for the usual parameters $n = 1{,}000$, $k = 10$. The most striking aspect of these figures is that $L(\xi)$ and $\gamma(\xi)$ have the *same functional form*. Fig. 3.49 displays this just for the one-dimensional case, but a similar result applies in two dimensions. This is quite different from the earlier results concerning relational graphs, for which $\gamma(\phi)$ decreased much more slowly than $L(\phi)$. Recall that this difference led directly to the class of small-world graphs. No such class can exist here because, by the stage the characteristic path length of the graph is comparable to its random-graph counterpart, so is its clustering coefficient. A second significant difference between uniform spatial and relational graphs is that if $L(\xi)$ and $\gamma(\xi)$ are recast in terms of ϕ, their functional forms resemble neither $L(\phi)$ nor $\gamma(\phi)$ for relational graphs (see Figs. 3.50 and 3.51). The same is true of the contractions parameter ψ (Figs. 3.52 and 3.53). Explaining these differences is a major task in Chapter 4.

Already, though, the basic idea is clear: whilst shortcuts and contractions may be possible in spatial graphs, they can never connect vertices

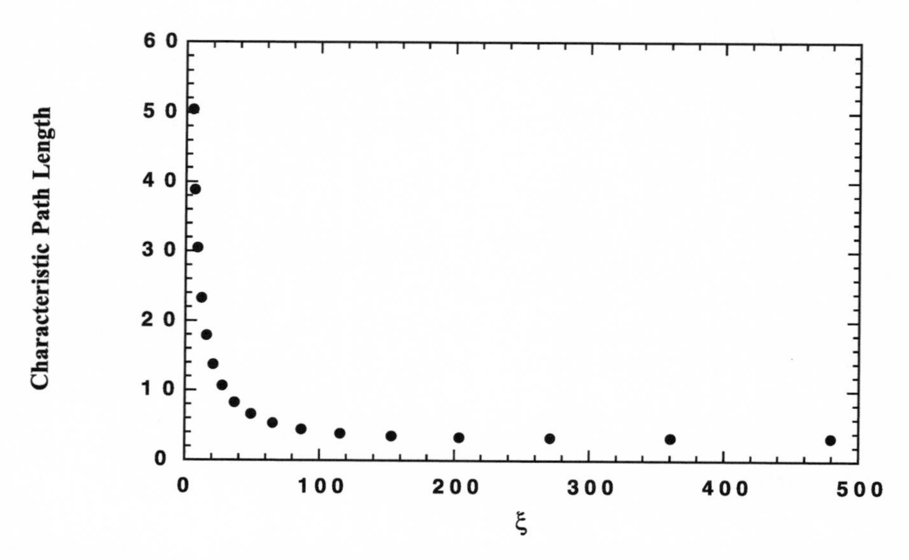

Figure 3.47 L vs. ξ for one-dimensional uniform spatial graphs ($n = 1{,}000$, $k = 10$).

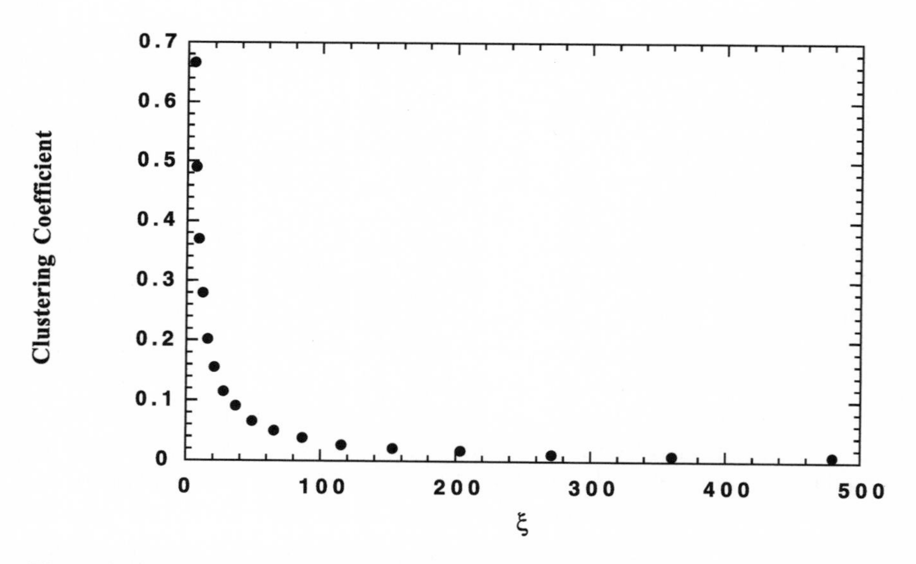

Figure 3.48 γ vs. ξ for one-dimensional uniform spatial graphs ($n = 1{,}000$, $k = 10$).

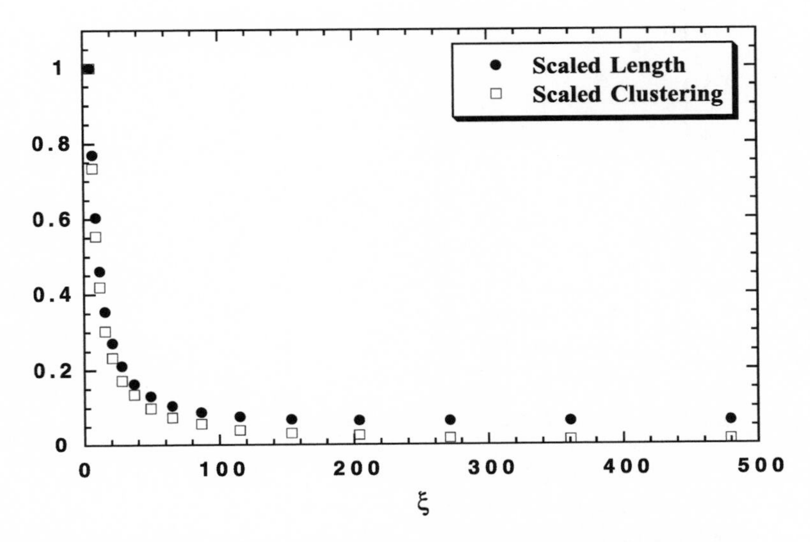

Figure 3.49 Scaled $L(\psi)$ and $\gamma(\psi)$ vs. ψ for uniform spatial graphs ($n = 1{,}000$, $k = 10$).

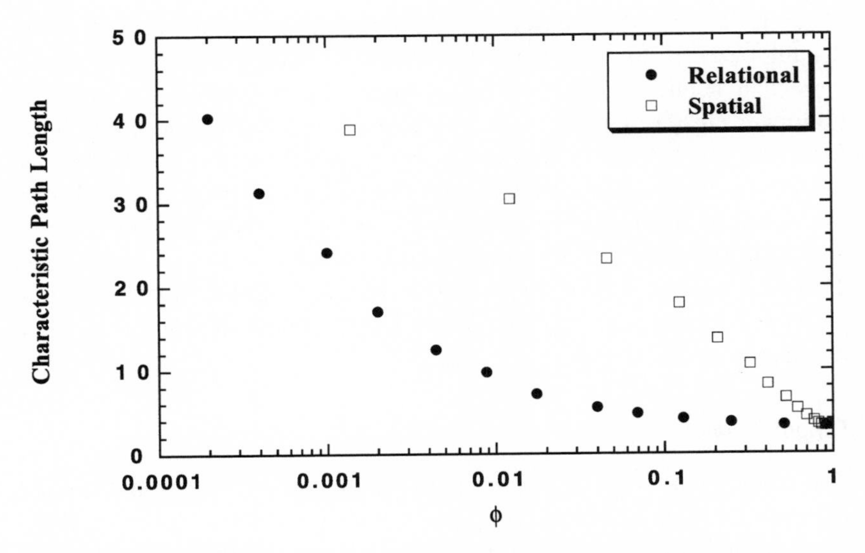

Figure 3.50 Comparison of $L(\phi)$ vs. ϕ for spatial and relational graphs ($n = 1{,}000$, $k = 10$).

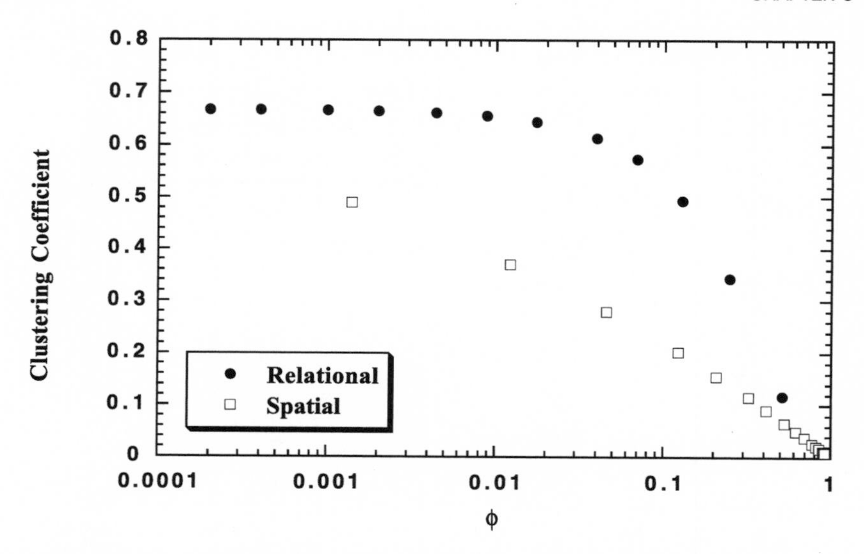

Figure 3.51 Comparison of $\gamma(\phi)$ vs. ϕ for spatial and relational graphs ($n = 1{,}000$, $k = 10$).

that were previously farther apart than the *external length scale* (as specified by ξ). That is, the connectivity is completely dominated by the cutoff of the probability distribution, so edges can never connect vertices from distant parts of the graph until that length scale is made sufficiently large that it encompasses the entire graph. By that stage, the graph is no longer clustered and has, in fact, approached its random-graph limit for clustering as well. In contrast, at very small ϕ, relational graphs may exhibit a small fraction of edges with ranges on a *global length scale* (on the order of the length of the graph), and these edges, whilst having very little effect on the clustering, have a dramatic impact on the length. This dual concept of two length scales (local and global) coexisting in a graph is the key to the explanations of length and clustering phenomena in Chapter 4.

Length Scaling

The other qualitative difference between spatial and relational graphs refers to their length-scaling properties. Recall that for relational graphs, logarithmic length scaling with respect to n appeared for even very small values of ϕ. It appears that, in the limit $n \to \infty$, an arbitarily small ϕ results in logarithmic length scaling. This does not happen with finite-cutoff spatial graphs. In contrast, as Figure 3.54 shows, the characteristic

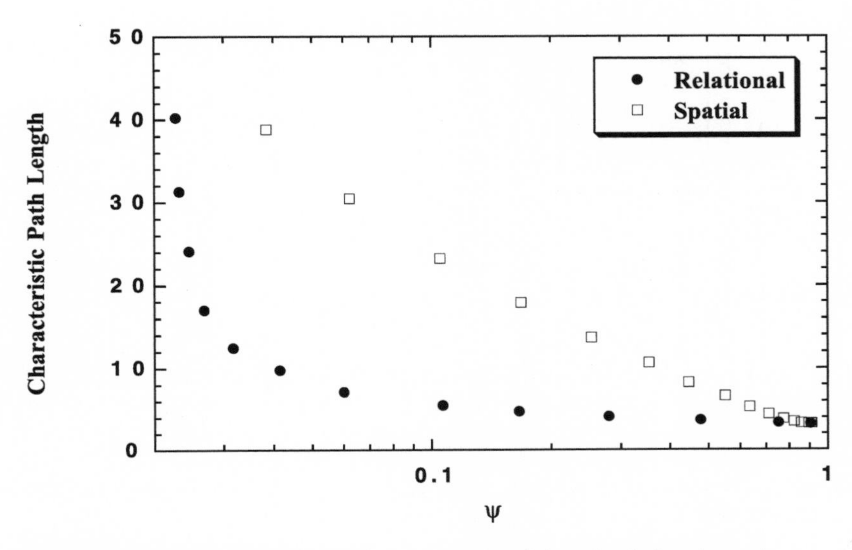

Figure 3.52 Comparison of $L(\psi)$ vs. ψ for spatial and relational graphs ($n = 1{,}000$, $k = 10$).

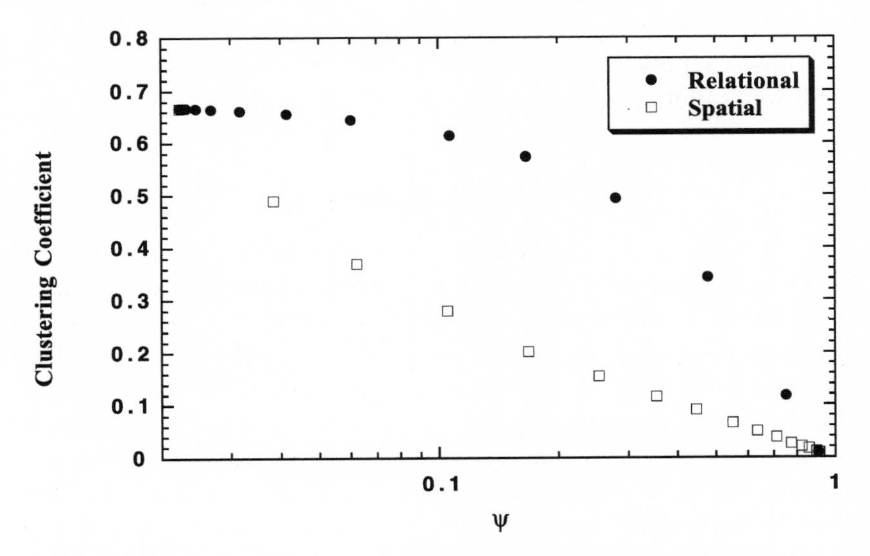

Figure 3.53 Comparison of $\gamma(\psi)$ vs. ψ for spatial and relational graphs ($n = 1{,}000$, $k = 10$).

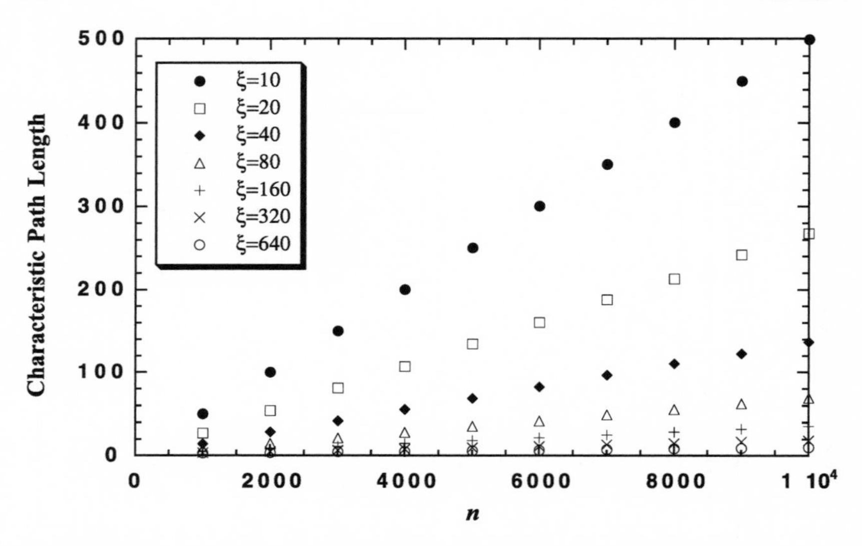

Figure 3.54 L vs. n for uniform spatial graphs with $k = 10$ for variables ξ. The length scaling is linear with respect to n for all $\xi = o(n)$.

path length of spatial graphs exhibits logarithmical scaling only when $\xi = O(n)$, that is, once the graphs have approached their random limit. This is another qualitative difference between the two models that is also dealt with in the next chapter.

3.2.2 Gaussian Spatial Graphs

The construction of a Gaussian spatial graph is almost identical to that of the uniform spatial graph: only the distribution has been changed. In other words, d coordinates are assigned according to a Gaussian rather than a uniform distribution in order to choose neighbours in $\mathbb{R}^d$. The standard yardstick of length in a Gaussian distribution is its standard deviation σ. As mentioned above, $\xi = 3\sigma$ is a measure of the half-width of the Gaussian distribution. Figures 3.55 and 3.56 show a comparison between Gaussian and uniform spatial graphs of $L(\xi)$ and $\gamma(\xi)$. The evidence here is that the details of the distribution used to build a spatial graph are unimportant compared with the parameter ξ, which appears to be the determining factor of the graphs' statistics.[15] Hence, from this point on, when the distribution of spatial graphs is not specified, it is assumed to be uniform.

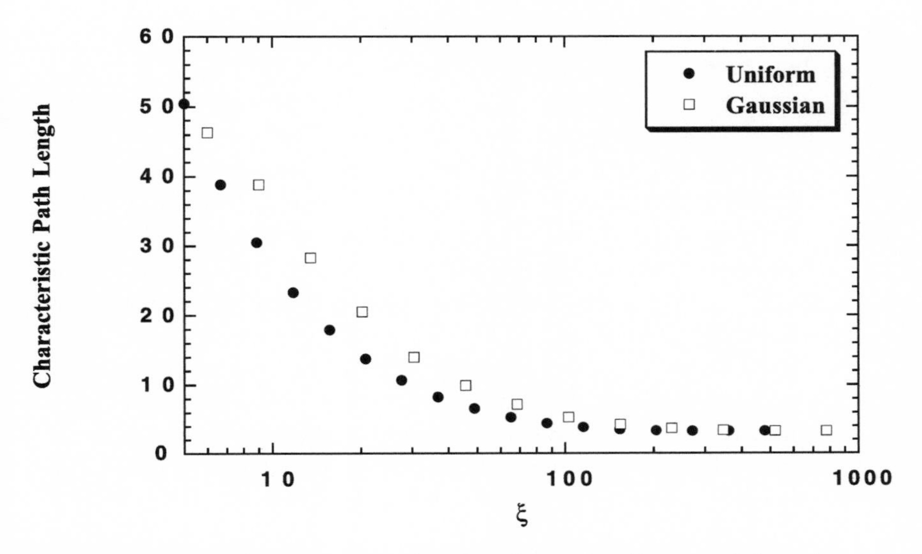

Figure 3.55 Comparison of L vs. ξ for Gaussian and uniform spatial graphs ($n = 1{,}000$, $k = 10$).

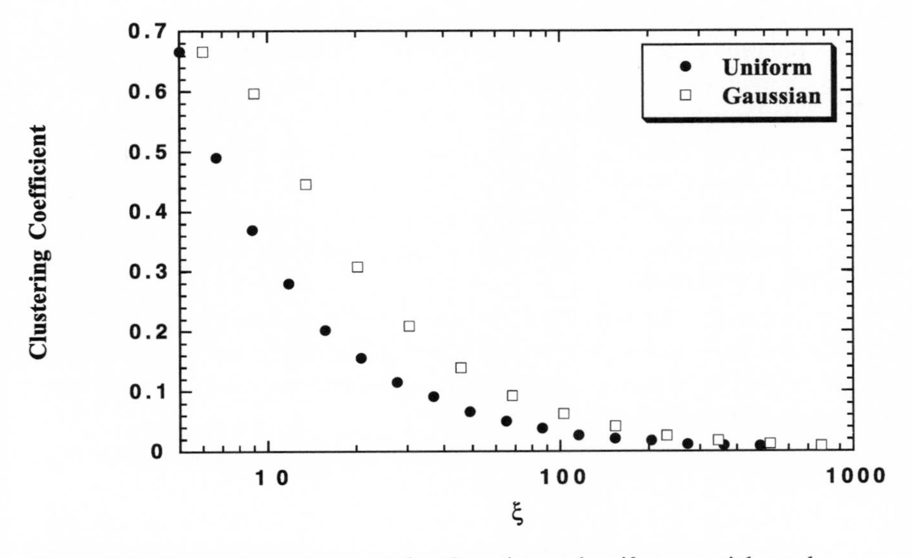

Figure 3.56 Comparison of γ vs. ξ for Gaussian and uniform spatial graphs ($n = 1{,}000$, $k = 10$).

3.3 MAIN POINTS IN REVIEW

1. Two broad classes of graph-construction algorithms are investigated, both of which interpolate between ordered and random limits. These classes are termed *relational* and *spatial* graphs, respectively. The probability of two vertices becoming connected in a relational graph depends only upon preexisiting connections. The corresponding probability in a spatial graph is a function of the Euclidean distance between the vertices. The only spatial graphs considered here are those whose defining probability distributions have a finite cutoff.
2. Different models of relational graphs can be unified by considering their dependence upon a model-independent parameter (ϕ), which quantifies the fraction of *shortcut edges*: edges that connect vertices that would otherwise be widely separated. A slightly more general mechanism is that of vertex contraction, quantified by the parameter ψ.
3. Different models of spatial graphs can be unified with respect to a different, model-independent parameter (ξ), which characterises an externally imposed, spatial length scale, beyond which edges cannot connect.
4. Relational graphs admit a particular class of graphs that exhibit characteristic path lengths approximately the same as equivalent random graphs (that is, $L \sim \ln(n)/\ln(k)$), but with much greater clustering. These graphs are called *small-world graphs*.
5. The spatial graphs considered (those with a finite cutoff) admit no such class, as clustering and characteristic path length possess the same functional dependence on the driving parameter (ξ) for that class of models.
6. In the limit $n \to \infty$, it appears that relational graphs with arbitrarily small ϕ (fraction of shortcut edges) will display logarithmic length scaling with respect to n and inverse logarithmic length scaling with respect to k.
7. One-dimensional spatial graphs exhibit logarithmic length scaling only in the limit where $\xi = O(n)$.
8. Relational graphs exhibit a peak in their structural complexity (as defined by the average variation in the range of their edges) for $\phi < 1$, which lies in the small-world regime.

4

Explanations and Ruminations

The simulations of the previous chapter have provided some intriguing results (and some intuition) concerning the structural nature of graphs that bridge the chasm between ordered and random domains, but little in the way of explanations. The goal of this chapter is to develop analytically penetrable models that approximate and explain the following phenomena:

1. The characteristic path length and clustering properties of
   a. ordered, highly clustered graphs and
   b. random graphs.
2. The functional form of the L- and γ-transitions between these two extremes for
   a. relational graphs and
   b. spatial graphs.
3. The length-scaling properties of relational and spatial graphs.
4. An explanation of why only relational graphs can be "small worlds" in the sense that locally they are highly clustered, but globally they exhibit the length and length-scaling properties of random graphs.

At this point it should be stressed that these explanations are not proofs in any mathematically rigorous sense. Rather, they are heuristic models, based on a number of drastic simplifications, which yield analytical approximations that show surprisingly good fit to the simulation data of Chapter 3.

4.1 GOING TO EXTREMES

The emphasis of this work is on graphs that combine properties of order and randomness in a fashion that is reminiscent of the "real world." In order to understand these, however, we first need to understand more about the graphs whose properties they are combining. This task can be approached by posing two questions. First, what is the most highly clustered graph for a given n and k, and what are its length and clustering properties? That is, *how big is the most clustered of all possible networks*? Second, what are the corresponding properties of the graph with the smallest possible characteristic length for a given n and k? That is, *how clustered is the smallest of all possible networks*?

The answers to these questions will provide a platform on which to build a model of partly ordered, partly random graphs that can help explain the length, clustering, and small-world properties observed in the graphs of the previous chapter.

4.1.1 The Connected-Caveman World

One obvious candidate for the most highly clustered graph possible is the *complete graph*, in which every vertex is adjacent to every other vertex. However, this construction automatically violates the sparseness condition, as it necessarily has $k = n - 1$. What is required, then, is a graph that is *globally sparse* but *locally dense*; that is, with $k \ll n$ and $\gamma \approx 1$.

A better solution is what might be termed the *caveman graph*, which consists of a number of fully connected clusters (or "caves") in which every member is adjacent with every other (Fig. 4.1). Every vertex has degree k, and so each cluster must consist of $n_{\text{local}} = k + 1$ vertices, and there must be $n_{\text{global}} = n/(k + 1)$ such clusters. Note also that all edges are part of multiple triads, so no edge is a shortcut. Hence the caveman graph satisfies the sparseness condition and still has $\gamma = 1$, but it fails to satisfy another essential condition: that it be connected. The best solution, then, appears to be a close approximation to the caveman graph that is both periodic and connected.

These conditions can be satisfied by a construction that is created by modifying the caveman graph in the manner illustrated in Figure 4.2. The resulting graph—the *connected-caveman graph* (Fig. 4.3)—exhibits a clustering coefficient that approaches that of the caveman graph in the large-n, k limit.

Clustering Coefficient

Recall (from Chapter 2) that the general expression for the clustering coefficient of a single vertex v is

$$\gamma_v = \frac{\text{total edges in } \Gamma(v)}{\text{total possible edges in } \Gamma(v)}$$

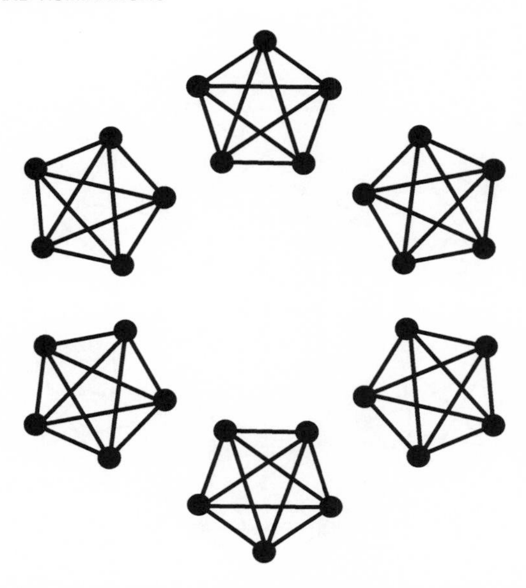

Figure 4.1 The caveman graph consists of $n/(k+1)$ isolated "caves," each of which is completely connected internally.

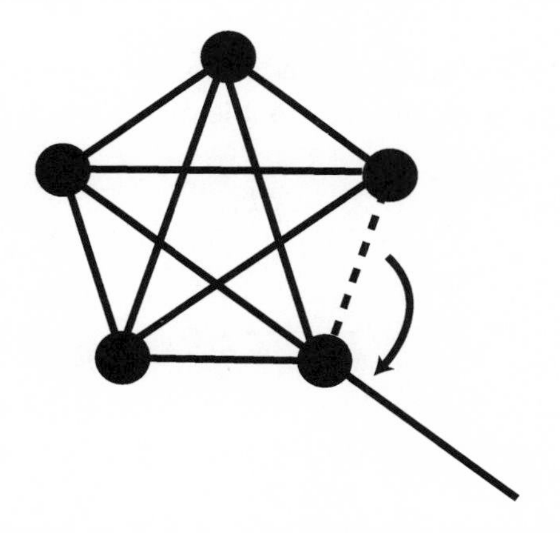

Figure 4.2 Procedure to connect the caveman graph.

and that γ is the average of γ_v over the entire graph. As the connected-caveman graph consists of identical clusters, then the problem reduces to that of averaging γ_v over all vertices in a single cluster, of which there are four kinds (see Fig. 4.3):

1. Type-a vertices ($k-2$ of these in each cluster, each one of which has $k_a = k$ neighbours). Every vertex in $\Gamma(a)$ is connected to every

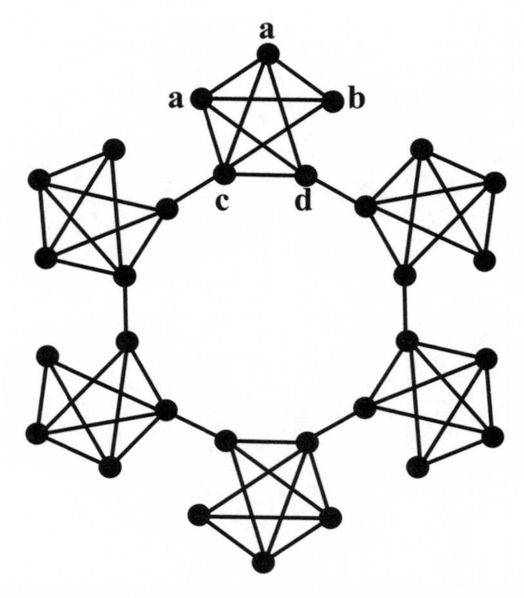

Figure 4.3 The connected caveman graph consists of a ring of caves, where the edges of the ring have been transferred from within the caves.

other, with the exception of edge (b, d). Hence there is one edge missing from an otherwise completely connected neighbourhood, implying that

$$\begin{aligned}\gamma_a &= \frac{2}{k(k-1)}\left(\frac{k(k-1)}{2} - 1\right) \\ &= 1 - \frac{2}{k(k-1)}.\end{aligned}$$

2. Type-b vertices (one of these in each cluster, with $k_b = k - 1$ neighbours). All vertices in $\Gamma(b)$ are connected to all others; hence
$$\gamma_b = 1.$$

3. Type-c vertices (one of these in each cluster, with $k_c = k + 1$ neighbours). The type-c vertex is connected to a type-d vertex from another cluster; hence it is missing k edges from its neighbourhood, in addition to the edge (b, d). So

$$\begin{aligned}\gamma_c &= \frac{2}{(k+1)k}\left(\frac{(k+1)k}{2} - (k+1)\right) \\ &= 1 - \frac{2}{k}.\end{aligned}$$

4. Type-d vertices (one of these in each cluster, with $k_d = k$ neighbours). The type-d vertex is connected to a type-c vertex from another cluster; hence it is missing $k-1$ edges from its neighbourhood. So

$$\begin{aligned}\gamma_d &= \frac{2}{k(k-1)}\left(\frac{k(k-1)}{2} - (k-1)\right)\\ &= 1 - \frac{2}{k}.\end{aligned}$$

A weighted average over all vertices in the cluster then yields

$$\begin{aligned}\gamma_{cc} &= \frac{1}{k+1}\left((k-2)\,\gamma_a + \gamma_b + \gamma_c + \gamma_d\right)\\ &= \frac{1}{k+1}\left[(k-2)\left(1 - \frac{2}{k(k-1)}\right) + 1 + 2\left(1 - \frac{2}{k}\right)\right]\\ &= \frac{1}{k+1}\left[(k+1) - \frac{6k-8}{k\,(k-1)}\right]\\ &= 1 - \frac{6}{k^2-1} + O\left(\frac{1}{k}\right)^3.\end{aligned}$$

This construction is not proven to be the most highly clustered graph that is consistent with the requirements of connectedness, sparseness, and periodicity. However, if it is not, then the graph that *is* the most clustered can only be more so by an amount of $O\left(1/k^2\right)$, which is negligible compared with the $O(1)$ difference between the connected-caveman graph and a random graph. Furthermore, as $n \to \infty$, k can be made arbitarily large without violating the sparseness condition, in which case $\gamma_{cc} \to 1$, making it effectively indistinguishable from the caveman graph.

The clustering coefficient can also be expressed in terms of the *effective degree* of the vertices. There are, in fact, two closely related but distinct variants of effective degree, both of which are relevant to calculating γ.

Definition 4.1.1. The *effective local degree* k_{local} is the average number of edges per vertex that have a range $r = 2$. That is, *local edges* are part of at least one triad, so k_{local} captures the number of edges belonging to a typical vertex that are *not* shortcuts.

Definition 4.1.2. The *effective clustering degree* k_{cluster} is the average number of vertices $u_{j \neq i} \in \Gamma(v)$ that each $u_i \in \Gamma(v)$ is connected to.[1] That is, k_{cluster} quantifies how many of v's *other* neighbours, each of v's neighbours is connected to.

Hence k_{local} is a property of the vertex v, and k_{cluster} is a property of v's neighbours. Because the introduction of shortcuts plays a crucial role in the models of this chapter, and because it is easier to analyse the effect of shortcuts in terms of vertices than it is in terms of neighbourhoods, it is convenient to reexpress the clustering coefficient in terms of k_{local} and k_{cluster}:

$$\begin{aligned}\gamma &= \frac{2}{k(k-1)} \frac{k_{\text{local}}(k_{\text{cluster}}-1)}{2} \\ &= \frac{k_{\text{local}}(k_{\text{cluster}}-1)}{k(k-1)}.\end{aligned} \tag{4.2}$$

In the connected-caveman model, k_{local} can be determined for each vertex type by inspection of Figure 4.3 and, averaging over all vertices,

$$\begin{aligned}k_{\text{local}_{cc}} &= \frac{1}{k+1}\left[(k-2)k + (k-1) + k + (k-1)\right] \\ &= \frac{k^2+k-2}{k+1} \\ &= k - \frac{2}{k+1}.\end{aligned} \tag{4.3}$$

Equating γ_{cc} (from Equation 4.1) with the right-hand side of Equation 4.2 and substituting Equation 4.3 for k_{local}, the effective clustering degree for the connected-caveman graph can be shown to be

$$k_{\text{cluster}_{cc}} = k - \frac{4}{k} + O\left(\frac{1}{k}\right)^3.$$

Characteristic Length

Because most pairs of vertices in the same cluster are adjacent with each other ($d(i, j) = 1$), and because, for $n \gg k$, most pairs of vertices consist of vertices from different clusters, the characteristic path length (L) for the connected-caveman graph is dominated by the shortest path length between *clusters*. In what follows, two distances are significant: the average distance between two vertices in the same cluster (d_{local}) and the average distance between vertices in different clusters (d_{global}). There are also two length scales, L_{local} and L_{global}, which are the characteristic path lengths within and between clusters, respectively. This may seem redundant, but the reason is that d_{global} is the average distance between *vertices* that are in different clusters, and L_{global} is the average distance between *clusters of vertices*, which, on the global scale, can themselves be treated like vertices. At the moment L_{local} *is* redundant, but it will be useful later. Meanwhile, it is easy to compute, as only two out of $(k+1)k/2$ pairs have $d(i,j) = 2$ and the rest have $d(i,j) = 1$ (see Fig. 4.2). Hence

$$d_{local} = L_{local} = \frac{2}{(k+1)k}\left[\left(\frac{(k+1)k}{2} - 2\right)\cdot 1 + 2\cdot 2\right] = 1 + \frac{4}{(k+1)k}, \tag{4.4}$$

which, for $k \gg 1$, simplifies to $d_{local} \approx 1$.

Each cluster can then be thought of as a *meta-vertex*, with its own internal length (L_{local}), implying that d_{global} is determined both by L_{local} and the *global length scale* L_{global}, which is just the characteristic length of a ring with $n_{global} = n/(k+1)$ and $k_{global} = 2$. Recall (from Chapter 2) that

$$L_{ring} = \frac{n(n+k-2)}{2k(n-1)}$$

$$\Rightarrow L_{global} = \frac{(\frac{n}{k+1})^2}{4(\frac{n}{k+1} - 1)}.$$

From Figure 4.4 we can see that a path from a vertex v in one cluster to a vertex u in another cluster consists of three components:

1. The number of edges traversed in order to get out of the cluster containing v (L_{local}).
2. The number of edges traversed moving between clusters. This will be two per cluster (one global edge and one local edge through the cluster) for each of ($L_{global} - 1$) clusters.
3. The number of edges traversed in order to get into the cluster containing u ($L_{local} + 1$). (The extra edge is the final step required to reach the cluster.)

Hence

$$\begin{aligned} d_{global} &= L_{local} + 2(L_{global} - 1) + (1 + L_{local}) \\ &= \frac{8}{k(k+1)} + \frac{(\frac{n}{k+1})^2}{2(\frac{n}{k+1} - 1)} + 1. \end{aligned} \tag{4.5}$$

For $n \gg k \gg 1$, this expression simplifies to $d_{global} \approx n/2(k+1)$.

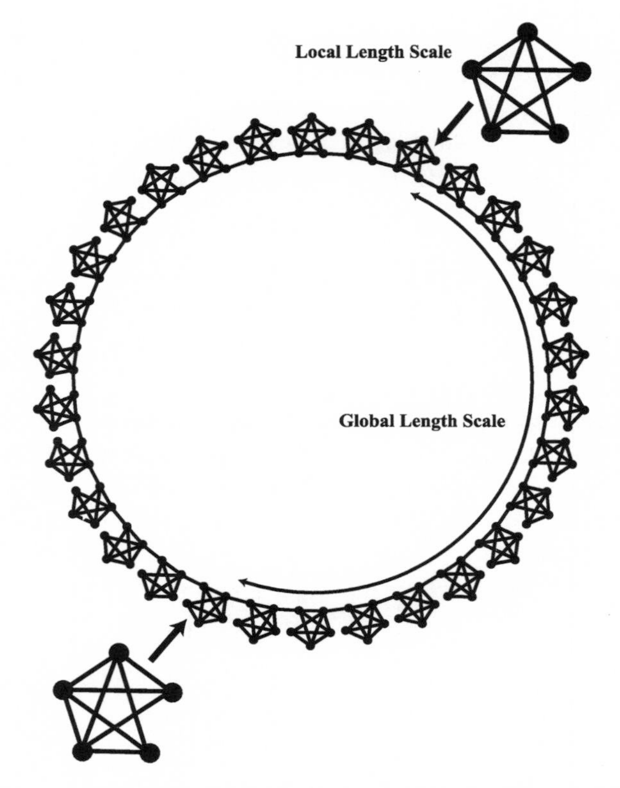

Figure 4.4 Graphical interpretation of the characteristic path-length calculation for vertices in different caves of the connected-caveman graph.

There are

$$\frac{(k+1)k}{2} \cdot \frac{n}{k+1} = \frac{n \cdot k}{2} = N_{\text{local}}$$

pairs of vertices that inhabit the same clusters and

$$\frac{n}{2(k+1)}\left[\left(\frac{n}{k+1} - 1\right) \cdot (k+1)^2\right] = \frac{n(n-k-1)}{2} = N_{\text{global}}$$

pairs that inhabit different clusters, the sum of which is $n(n-1)/2 = N$. Hence the average distance between all pairs of vertices is

$$\begin{aligned} L_{cc} &= \frac{1}{N}\left[N_{\text{local}} \cdot d_{\text{local}} + N_{\text{global}} \cdot d_{\text{global}}\right] \\ &\approx \frac{2}{n(n-1)}\left[\frac{n \cdot k}{2} \cdot 1 + \frac{n(n-k-1)}{2} \cdot \frac{n}{2(k+1)}\right] \end{aligned}$$

$$= \frac{k}{n-1} + \frac{n(n-k-1)}{2(k+1)(n-1)}$$

$$\approx \frac{n}{2(k+1)} \quad (n \gg k \gg 1). \tag{4.6}$$

Note that this is similar to the expression for the characteristic length of a 1-lattice with n vertices and degree k. This is equivalent to the statement that L for the caveman graph is dominated by d_{global}, the distance between vertices in different clusters. This observation will be useful in Section 4.2 in order to approximate the characteristic length of the entire graph, as a function of ϕ.

4.1.2 Moore Graphs as Approximate Random Graphs

Turning to the second question posed above—What is the characteristic length and clustering properties of the smallest of all possible networks?—a lower bound on L is not difficult to construct, if one assumes that all vertices have the same degree. The relevant construction is known as a *Moore Graph* (Bollobás 1985), which is a *perfectly expanding graph* in the sense that every vertex is adjacent with precisely k vertices, none of which are adjacent with each other. Locally, a Moore Graph must look like Figure 4.5, but understandably there may be problems at the boundaries of a finite graph where all the edges must fit together in just such a fashion that neither perfect regularity nor perfect expansion are compromised. In fact, it can be shown that Moore Graphs are not attainable for most n and k (Cerf et al. 1974), and in a few other cases their construction remains an open problem (Chapter 2 of Bollobás 1985). Nevertheless, they *are* guaranteed to place a lower bound on L, at least for graphs of regular degree.

Characteristic Length

The characteristic path length of this theoretical lower bound is straightforward to calculate. First note (from Fig. 4.5) that at distance d from any vertex v (locally the graph looks identical from any vertex), $k(k-1)^{d-1}$ vertices can be reached. Hence when the most distant parts of the graph are reached, D steps from v, the number of vertices already included is

$$S = \sum_{d=1}^{D-1} k(k-1)^{d-1}.$$

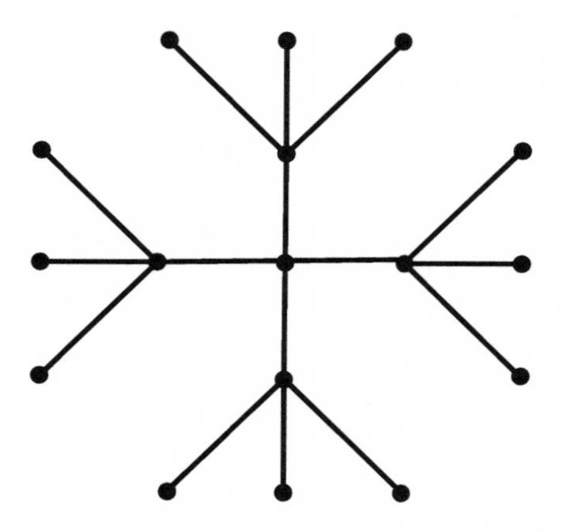

Figure 4.5 Local view around a vertex in a Moore Graph.

The remaining $(n - S - 1)$ vertices are naturally at a distance D from v, hence the sum of all distances from v is

$$\sum_i L_{v,i} = \sum_{d=1}^{D-1} d \cdot k(k-1)^{d-1} + (n - S - 1) \cdot D.$$

Averaging over the $(n - 1)$ vertices (besides v) in the graph, and noting that L_v is the same for all v,

$$L_{\mathrm{M}} = \frac{1}{n-1}\left[\sum_{d=1}^{D-1} d \cdot k(k-1)^{d-1} + (n - S - 1) \cdot D\right],$$

where D is the *diameter* of the graph. Now, using the summation formulae

$$\sum_{i=1}^{I} r^{i-1} = \frac{r^I - 1}{r - 1}$$

and

$$\sum_{i=1}^{I} i \cdot r^{i-1} = \frac{nr^{I+1} - (I+1)r^I + 1}{(r-1)^2} \qquad (r \neq 2),$$

then, for $k > 2$,

$$L_{\mathrm{M}} = \frac{1}{n-1}\left(k\left[\frac{(D-1)(k-1)^D - D(k-1)^{D-1} + 1}{(k-2)^2}\right] + D\left[n - \frac{(k-1)^{D-1} - 1}{k-2} - 1\right]\right), \qquad (4.7)$$

which can be reexpressed as

$$L_M = D - \frac{k(k-1)^D}{(n-1)(k-2)^2} + \frac{k(D(k-2)+1)}{(n-1)(k-2)^2} \quad (k > 2), \tag{4.8}$$

where D can be approximated as

$$S = \sum_{d=1}^{D-1} k(k-1)^{d-1} \leq n-1$$

$$\Rightarrow k\frac{(k-1)^{D-1}-1}{k-2} \leq n-1$$

$$\Rightarrow D \leq \frac{\ln\left[\frac{(k-2)}{k}(n-1)+1\right]}{\ln(k-1)} + 1$$

$$\Rightarrow D = \left\lfloor \frac{\ln\left[\frac{(k-2)}{k}(n-1)+1\right]}{\ln(k-1)} + 1 \right\rfloor \quad (k > 2). \tag{4.9}$$

The condition $k > 2$ is necessary if the graph is to expand, that is, if an exponentially growing number of vertices is to be reached at each step away from v. If $k = 2$, then a Moore Graph is simply a ring, in which case (for even n)

$$L_M = \frac{n^2}{4(n-1)} \quad (k = 2). \tag{4.10}$$

This expression is also useful for approximating the characteristic length of a random graph, which is the limit to which all the one-parameter models are designed to converge for large values of the relevant parameter. No closed-form expression for the expected length of a random graph has been obtained, although Bollobás (1985, chapter 10) has bounded the diameter of random graphs, regular or otherwise, and Schneck et al. (1997) have considered the sum of all distances in a random graph.

Naturally a random graph is *not* a Moore Graph as it is neither perfectly regular in degree nor perfectly expanding. However, the redundancy that results from neighbours of any vertex v connecting randomly to other neighbours of v is only $O(k/n)$. Because this is true for every vertex, independent of any other, then the distance degree sequence will still grow exponentially (with respect to the number of steps away from v), and so we might expect that the L will change very little as a result of a small amount of clustering. In fact, this hypothesis can be tested by comparing the length of a Moore Graph to a *pseudo-random* graph, whose construction is defined as follows:

1. Construct a ring of n vertices (i.e., with $k = 2$).
2. Add edges to this ring at random, until there is a total of M edges, including the n edges used to form the ring. Then $k = 2M/n$, by definition.

Hence for $k = 2$, a pseudo-random graph is precisely a ring, and for $k > 2$, L decreases rapidly. Similar constructions have been studied by Chung (1986), who bounds the least possible diameter of cycles augmented in any fashion with additional edges, and Bollobás and Chung (1988), who consider a *random matching* on a cycle, in which each vertex on the cycle is assigned one (and only one) additional neighbour at random. They prove that the diameter of such a graph (which effectively has $k = 3$) is about $\log_2(n)$: close to the minimum possible diameter for *any* regular $k = 3$ graph.

Figure 4.6 shows that at $k \approx \ln(n)$, the characteristic length for the pseudo-random construction (L_{pseudo}) is indistinguishable from L_{random} for a random graph with the same n and k. Figure 4.7 shows a comparison between L_{pseudo} and L_{M} for $n = 1{,}000$, $2 \leq k \leq 12$. From these figures, it seems that the Moore Graph formula yields a good approximation to a pseudo-random graph, which in turn (for the relevant parameter range) approximates a random graph. It seems reasonable then to use a Moore Graph construction to approximate the characteristic path length of a random graph for $k \gtrsim \ln(n) \gg 1$.

Clustering Coefficient

Obviously one consequence of a perfectly expanding Moore Graph is that it contains no triads; hence no edges exist in $\Gamma(v)$ for any vertex v. That is, all edges are "reaching out into new territory" rather than connecting to other vertices in the same neighbourhood. Necessarily then,

$$\gamma_{\text{M}} = 0. \tag{4.11}$$

As pointed out above, this is not quite true for a random graph, where some triads may occur through random chance. In fact, the expected clustering coefficient for a k-regular random graph is

$$\gamma_{\text{random}} = \frac{k-1}{n} \approx \frac{k}{n}, \tag{4.12}$$

which becomes vanishingly small for $n \gg k$, at which point Moore Graphs and random graphs are again observed to be similar.

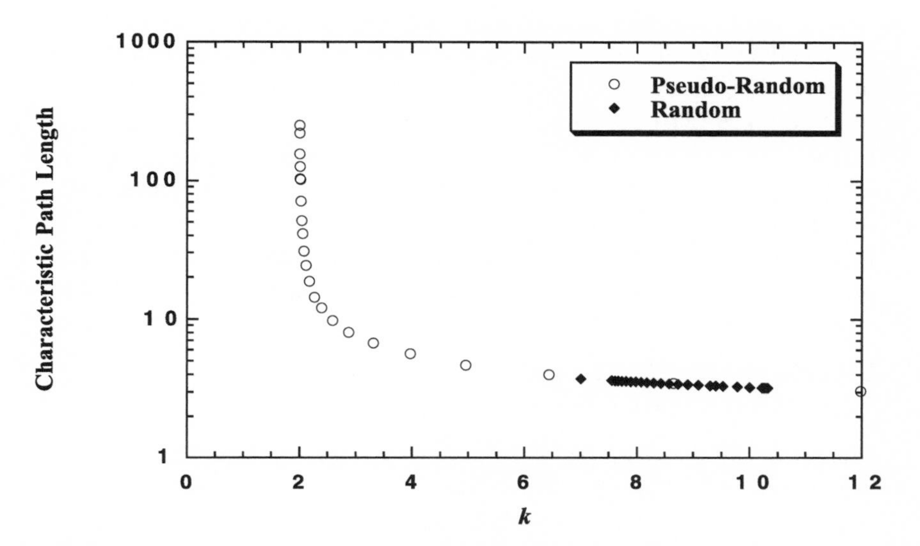

Figure 4.6 Comparison of L vs. k for pseudo-random graph (i.e., random edges added to a ring substrate) and a random graph (with $k \gtrsim \ln(n)$). Note that pseudo-random and random graphs agree over the range of k for which the random graph is connected ($k \gtrsim \ln(n)$).

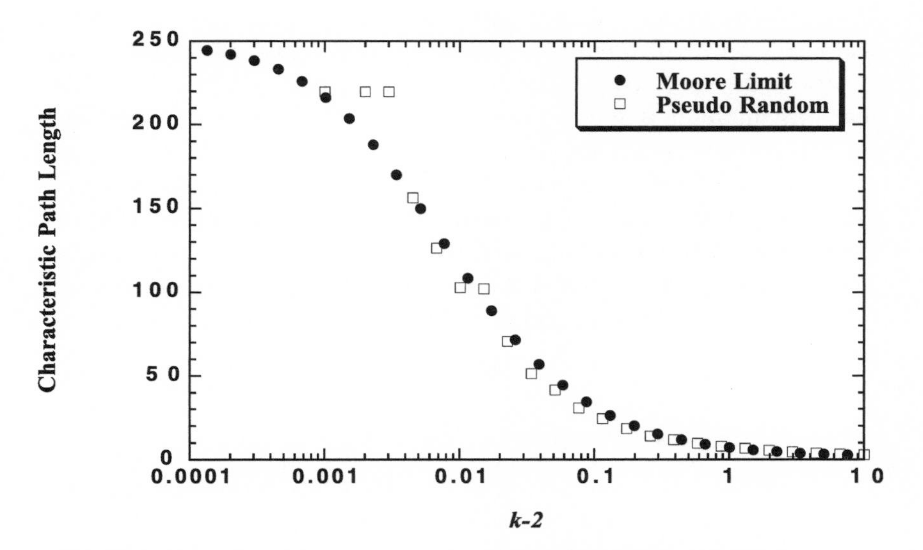

Figure 4.7 Comparison of L vs. $(k - 2)$ for the pseudo-random graph and a theoretical Moore Graph.

We are now in a position to define a *small-world graph* solely in terms of n and k. In Chapter 3, small-world graphs were defined in terms of γ for the most highly clustered graphs in a one-parameter family (that is, for $\phi = 0$) and close to that of the least-clustered graphs (that is, for $\phi \approx 1$). This definition, however, is satisfactory only when we have available an entire family of graphs so that we may compare the statistics of any one graph with those of graphs at the extremes of the family. Now, however, a small-world graph can be defined independently of any particular model or family of graphs, that is, solely in terms of an equivalent random graph.

Definition 4.1.3. A *small-world graph* is a graph with n vertices and average degree k that exhibits $L \approx L_{\text{random}}(n, k)$, but $\gamma \gg \gamma_{\text{random}} \approx k/n$.

Thus, for any given graph, it is possible to determine whether or not it is a small-world graph *without knowing anything of its construction* or requiring it to be a part of family of graphs. This feature will be especially advantageous in dealing with "real" graphs in Chapter 5.

4.2 TRANSITIONS IN RELATIONAL GRAPHS

The L and γ formulae of the previous section might lead one to contend that *highly clustered graphs necessarily have large L and graphs with small L necessarily have small* γ. The results of Chapter 3, of course, suggest that this intuition is sometimes correct (for spatial graphs with finite cutoffs) and sometimes not (for relational graphs). What the connected-caveman and Moore Graph models can add to this picture is an *analytical formulation* of the transition between ordered and random graphs that is driven by the introduction of shortcuts to the connected-caveman graph. The advantage of an analytical approach is that it can make more precise the conditions under which partly ordered, partly random graphs can be small worlds.

4.2.1 Local and Global Length Scales

The key concept to be utilised in this model is that of *dual length scales* that can be thought of as existing simultaneously in the graph: a *local length scale* (L_{local}) and a *global length scale* (L_{global}).[2] This idea has already arisen in the context of the caveman graph, where L_{local} characterised the length of a single cluster and L_{global} characterised the length

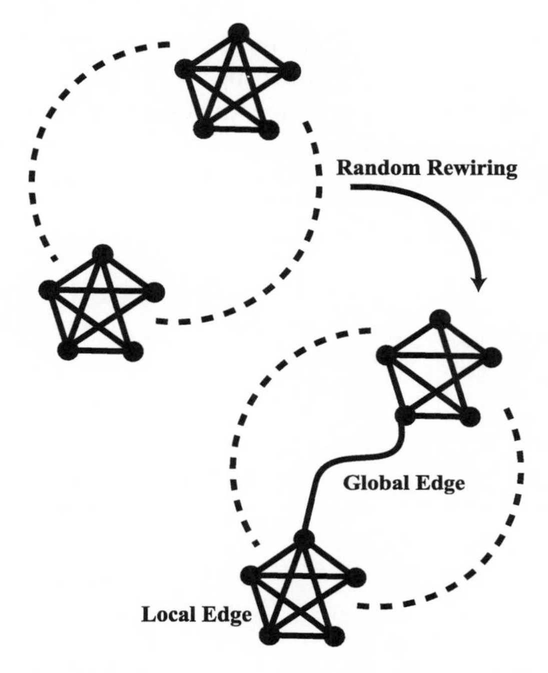

Figure 4.8 The transition between the connected-caveman graph and a random graph in the relational-graph model results from the transfer of edges from the local to the global scale.

of a ring of clusters. In fact, the clusters can be treated as *meta-vertices* with their own internal length scale. But what happens to these two length scales when edges that previously connected two vertices in the same cluster are removed and replaced by edges that connect vertices between clusters? That is, what happens when shortcuts are introduced into the connected-caveman graph via random rewiring? This is equivalent to shifting edges from the local to the global length scale (see Fig. 4.8). Every shortcut is equivalent to an edge removed from a cluster that depletes that cluster, causing L_{local} to increase; simultaneously an edge is added to the ring of meta-vertices at the global scale, causing L_{global} to decrease. As we saw in Section 4.1.1, the overall characteristic path length is dominated by d_{global}—the average distance between vertices in separate clusters—so it is possible to make the approximation $L \approx d_{\text{global}}$. But Equation 4.5 assumes that each cluster can be traversed by a single edge (the edge between the type-c and type-d vertices in Fig. 4.3). This is true for the connected-caveman graph (because it was specifically built that way) but will not be true in general, especially as edges are shifted randomly from the local to global scales. Hence the expression for d_{global} must be modified using the *variable* length L_{local} as a measure of the number of edges traversed in "moving through" a cluster. Specifically, from Equation 4.5 and assuming $L \approx d_{\text{global}}$:

$$\begin{aligned} L &\approx L_{\text{local}} + (1 + L_{\text{local}})\left(L_{\text{global}} - 1\right) + 1 + L_{\text{local}} \\ &= L_{\text{local}} + L_{\text{global}}\,(1 + L_{\text{local}})\,. \end{aligned} \tag{4.13}$$

4.2.2 Length and Length Scaling

The next important concept concerns the calculation of L_{local} and L_{global}. Initially each local cluster approximates a fully connected (*complete*) subgraph, and the global scale is exactly a ring. However, noting that a random graph with $k = (n - 1)$ is *also* a complete graph and, at the other extreme of connectivity, a random graph with $k = 2$ is *also* a ring, then both L_{local} and L_{global} can be approximated by L_{M} with the appropriate parameters. At the local scale, before any shortcuts are introduced, the connected-caveman model yields $k_{\text{local}} = k - 2/(k + 1)$ and $n_{\text{local}} = k + 1$, and (at the global scale) $k_{\text{global}} = 2$ and $n_{\text{global}} = n/(k + 1)$. As shortcuts are introduced through random rewiring (such that the total number of edges in the graph is fixed at $(k \cdot n)/2$) and *assuming* that local edges are converted to global edges at a rate linearly dependent on ϕ, then

$$\begin{aligned} k_{\text{local}} &= (1 - \phi)\left(k - \frac{2}{k+1}\right), \\ n_{\text{local}} &= k + 1 \\ k_{\text{global}} &= 2\,(1 - \phi) + k\,(k + 1)\,\phi, \\ n_{\text{global}} &= \frac{n}{k+1}. \end{aligned} \tag{4.14}$$

Using these parameters, the characteristic path length of a relational graph (L_r) can be expressed as a function of ϕ:

$$\begin{aligned} L_r &= L_{\text{M}}\,(n_{\text{local}}, k_{\text{local}}\,(\phi)) + L_{\text{M}}\left(n_{\text{global}}, k_{\text{global}}\,(\phi)\right) \\ &\times [L_{\text{M}}\,(n_{\text{local}}, k_{\text{local}}\,(\phi)) + 1] \\ &= L_{\text{M}}\left(k + 1, (1 - \phi)\left(k - \frac{2}{k+1}\right)\right) \end{aligned}$$

$$+ L_M\left(\frac{n}{k+1}, 2(1-\phi) + k(k+1)\phi\right)$$
$$\times \left[L_M\left(k+1, (1-\phi)\left(k - \frac{2}{k+1}\right)\right) + 1\right], \qquad (4.15)$$

where L_M is evaluated using Equation 4.8. For $\phi = 0$, the system is precisely the caveman graph, and indeed Equation 4.15 reduces to the expression for L_{cc} in the limit $\phi = 0$, where Equation 4.10 is used for the length of a Moore Graph with $k = 2$.

At this point it should be emphasised just how crude the above expression is. In constructing this approximation, the following are assumed:

1. Only two length scales exist in the problem.
2. Edges are moved from local to global scales at a rate linear in ϕ and uniformly across the entire graph.
3. All vertices have the same degree k at all values of ϕ.
4. The characteristic path length of a random graph is the same as that of a (probably unrealisable) Moore Graph with the same n and k.

In reality, none of these assumptions are likely to hold, and this will become apparent at large ϕ. In fact, as ϕ increases, the two length scales determined by Equation 4.8 (with the parameters of 4.14) will converge until they are the same. This collision of local and global length scales defines a value of ϕ beyond which the approximation is invalid and L_{local} blows up. Nevertheless much insight can be gained from this approach.

A particularly important consequence of Equation 4.15 concerns the length-scaling properties of relational graphs. The numerical results of Chapter 3 implied that relational graphs would display logarithmic length scaling with respect to n for $\phi \geq \phi_*$, as long as $n_{\min} > 2/(k\phi_*)$. Hence, in the limit $n \to \infty$, logarithmic scaling should occur for *arbitrarily small* ϕ. Inspection of Equations 4.8, 4.9, 4.10, and 4.15 reveal that when $k_{\text{global}} = 2$ (that is, $\phi = 0$), $L_r \propto n/k$, but that for $k_{\text{global}} > 2$ (that is, $\phi > 0$), even infinitesimally, $L_r \propto \ln(n)/\ln(k)$, in agreement with the numerical results.

4.2.3 Clustering Coefficient

In addition to predicting the characteristic path length and length-scaling properties of relational graphs, this model also yields an analytical approximation to $\gamma(\phi)$, as ϕ interpolates between the connected-caveman and random extremes. This is where the formulation of γ in terms of

k_{local} and k_{cluster} becomes useful. In calculating $\gamma(\phi)$, it is reasonable to assume not only that k_{local} decreases linearly with increasing ϕ, but that k_{cluster} does also. Substituting the relevant expressions into Equation 4.2, these approximations yield the following expression for γ of a relational graph:

$$\begin{aligned}\gamma_r(\phi) &= \frac{(1-\phi)k_{\text{local}}[(1-\phi)k_{\text{cluster}}-1]}{k(k-1)} \\ &= \frac{(1-\phi)(k-\frac{2}{k+1})[(1-\phi)(k-\frac{4}{k})-1]}{k(k-1)},\end{aligned}$$

and neglecting $O(1/k)^3$ terms,

$$\gamma_r(\phi) = 1 - 2\phi + \phi^2 - (\phi - \phi^2)\frac{1}{k} + (11\phi - 5\phi^2 - 6)\frac{1}{k^2} + O\left(\frac{1}{k}\right)^3. \tag{4.16}$$

Inspection of Equation 4.16 shows that, for small ϕ and large k, $\gamma(\phi) \sim (1 - 2\phi)$, confirming that, while the first few shortcuts have a highly nonlinear impact on $L(\phi)$, their effect on $\gamma(\phi)$ is only linear.

4.2.4 Contractions

Finally, the same model can also be modified to predict the functional dependence of contractions (defined in Chapter 3) on shortcuts and thus the dependence of L and γ on ψ, as opposed to ϕ. Recall from Chapter 3 that shortcuts and contractions, although defined independently, are intimately related. The motivation for contractions came from social networks that, rather than having one large, highly interconnected group of friends and a number of isolated (shortcut) relationships, consist of circles of friends, within each of which many relationships exist, but between which connections are relatively rare. Such a formation does not necessarily involve any shortcuts but will generally result in contractions as long as the individual *groups* of friends would be widely separated, were it not for a single common member.[3] In the context of the relational-graph model, consider a situation in which shortcuts are not made individually, but in *bundles* where all edges in the bundle connect a single vertex in one cluster to multiple vertices in another *single* cluster (see Fig. 4.9). Hence none of the edges are shortcuts because of

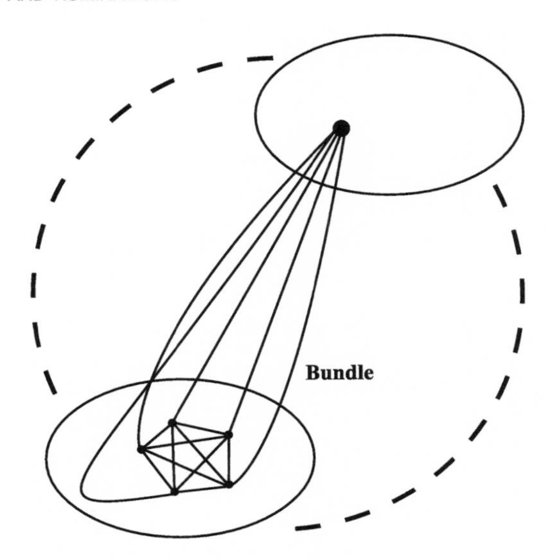

Figure 4.9 Edges connecting a single vertex in one cave to vertices that are all in another, single cave form a bundle. Bundles result in contractions but no shortcuts.

the high intraconnectedness of the receiving cluster, even though each edge connects vertices in different clusters. An analogy for this may be found in the marriage between two families. Initially two strangers meet and a shortcut connection forms between their two, possibly widely separated, families. However, their relationship necessitates each person's meeting additional members of the other's family. At no stage in the relationship, however, need any other member of one family ever meet any other member of the other family directly (especially if the couple elopes). Hence their relationship is no longer a shortcut (because multiple triads have been formed), but each party in the relationship is responsible for numerous contractions between members of the two families.

Contractions can be calculated as a function of shortcuts by considering the number of contractions that result when a vertex v makes b shortcuts to the same cluster in a *shortcut bundle* of size b. If $(k - b)$ vertices remain in the *main group* of $\Gamma(v)$ then $b(k - b)$ contractions are created. Furthermore, if v has n_b such bundles, then there are $n_b b(k - n_b \cdot b)$ contractions between all the bundles and the main group and a further $n_b/2[(n_b - 1)b^2]$ contractions between the bundles themselves (assuming here that the bundles are, themselves, widely separated), leading to

$$\begin{aligned}&\text{Total contractions from shortcut bundles}\\&= n_b b(k - n_b b) + \frac{n_b}{2}(n_b - 1)b^2.\end{aligned}$$

But these bundles consist of many individual shortcuts so $n_b b = \phi k$, and the number of contractions can be expressed in terms of ϕ and b alone:

$$\begin{aligned}&\text{Total contractions from shortcut bundles}\\&= k\phi(k - k\phi) + \frac{(k\phi)^2}{2} - \frac{k\phi b}{2}\\&= \frac{k(2k - b)\phi - (k\phi)^2}{2}.\end{aligned}$$

Dividing by $k(k-1)/2$—the total number of pairs of vertices in $\Gamma(v)$—the fraction of contractions for a relational graph (ψ_r) as a function of shortcuts is

$$\psi_r(\phi; b) = \frac{(2k - b)\phi - k\phi^2}{k - 1} + \psi_{\text{substrate}}, \tag{4.17}$$

where $\psi_{\text{substrate}}$ is an extra term that accounts for the contractions present before any shortcuts are made (that is, for $\phi = 0$). For the caveman graph, each type-*c* vertex creates k contractions, and each type-*d* vertex creates $(k - 1)$ contractions. Hence, averaging over all vertices in a cluster,

$$\psi_{\text{substrate}} = \psi_{\text{caveman}} = \frac{2k - 1}{k + 1} \cdot \frac{2}{k(k - 1)} = \frac{2(2k - 1)}{k(k^2 - 1)}.$$

This term does not affect the functional form of ψ but is important in order to make quantitative comparisons with numerical results.

4.2.5 Results and Comparisons with β-Model

The model presented above describes a particular means of traversing between two artifical constructions: the connected-caveman graph and the Moore Graph. The connected-caveman graph was chosen specifically because it is conceptually easy to motivate and naturally divides into two length scales, which simplifies its analysis. However, in order to make a *quantitative* comparison between the output of the resulting model and the numerical results of Chapter 3, for the same n and k, the formulae for $L(\phi)$, $\gamma(\phi)$, and $\psi(\phi)$ must be reexpressed in terms of the appropriate starting point of the random-rewiring procedure. Because the β-model is the simplest of the two relational-graph models, it makes sense to redo the formulae in terms of a 1-lattice.

It turns out that this is not difficult to do, allowing for one assumption: that L_{local} for a 1-lattice can be treated as L_{local} for the connected-caveman graph, but where k_{local} is reduced appropriately to account for the fact that nearby vertices on a 1-lattice are not as densely interconnected as those of the caveman clusters. This might seem like something of a stretch at first, as a 1-lattice is certainly *not* a ring of clusters joined by single edges. However, it *is* true that the neighborhood of a vertex $\Gamma(v)$ defines a natural length scale, in that every vertex within that neighbourhood can be reached in one step from v. Hence the 1-lattice model can be forced into the connected-caveman mold by thinking of the distance between any two vertices in the lattice as divided into regions of length $L(\Gamma(v)) = L_{\text{local}}$ and realising that, as with the caveman graph, there are $n/(k+1)$, of these and so there will be (on average) L_{global} of them separating any two vertices (see Fig. 4.10). Thus the characteristic length of a sparsely connected graph is dominated by the distances between pairs of vertices that are not in the same neighbourhood, and these distances are, in turn, determined roughly by the *number of neighbourhoods* that lie between each pair vertices.

Accepting this, then Equation 4.15 for L_r (obtained using the caveman substrate) remains *unchanged* for the 1-lattice substrate. Using this approximation, L_r from Equation 4.15 can be compared with the numerical output of the β-model from Chapter 3 for the usual parameters of $n = 1{,}000$, $k = 10$ (Fig. 4.11). The agreement between theory and numerics is striking and remains so even for a much larger graph with $n = 20{,}000$, $k = 10$ (Fig. 4.12).

In contrast with L_r, the calculation for γ_r is inherently local and so must account for the different local structure of the 1-lattice. Unlike the connected-caveman graph, all vertices in the 1-lattice have identical neighbourhoods, so the edges that contribute to clustering can be enumerated as follows:

$$\begin{aligned} k_{\text{cluster}} &= \frac{2}{k}\sum_{i=1}^{\frac{k}{2}}(k-i) \\ &= \frac{2}{k}\left[\frac{k^2}{2} - \frac{k}{4}\left(\frac{k}{2}+1\right)\right] \\ &= \frac{3}{4}\left(k - \frac{2}{3}\right). \end{aligned}$$

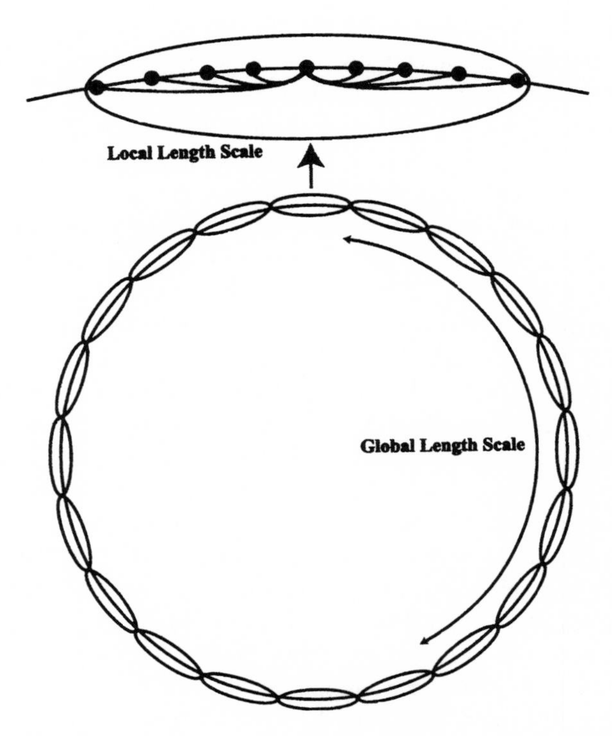

Figure 4.10 Schematic of the characteristic path-length calculation for the β-graph model.

Assuming, as before, that local edges are converted to global edges at a rate linearly dependent upon ϕ, and substituting for the corresponding variables in Equation 4.2 (where now $k_{\text{local}} = k$):

$$\begin{aligned}\gamma_\beta(\phi) &= \frac{k_{\text{cluster}} - 1}{k - 1} \\ &= \frac{\frac{3}{4}(1-\phi)^2\left(k - \frac{2}{3}\right) - (1-\phi)}{k-1}.\end{aligned} \tag{4.18}$$

Figures 4.13 and 4.14 show the resulting comparisons between the analytical and numerical versions of γ_β. Surprisingly for such a simplistic approximation, the agreement is close for a wide interval of ϕ and excellent for small ϕ, where the clustering is still dominated by the $\phi = 0$ limit.

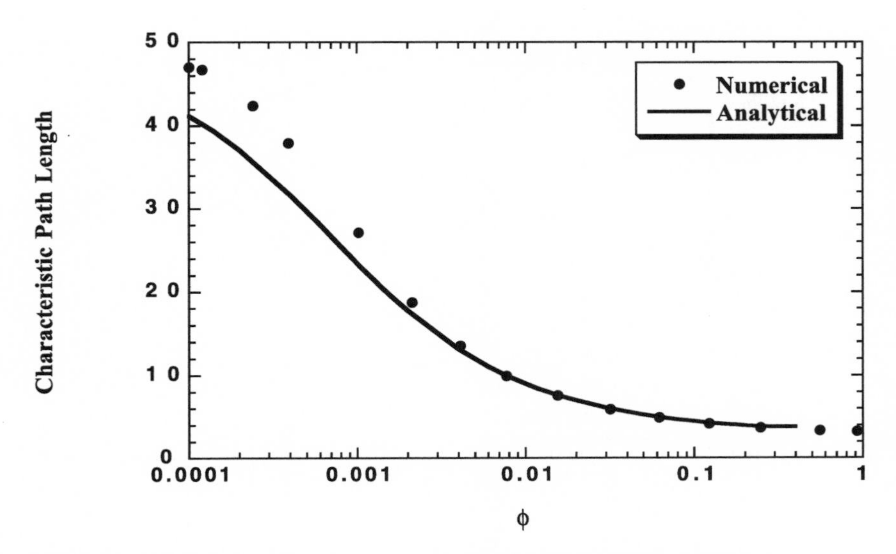

Figure 4.11 Comparison of analytical and numerical $L(\phi)$ for β-graphs ($n = 1{,}000$, $k = 10$).

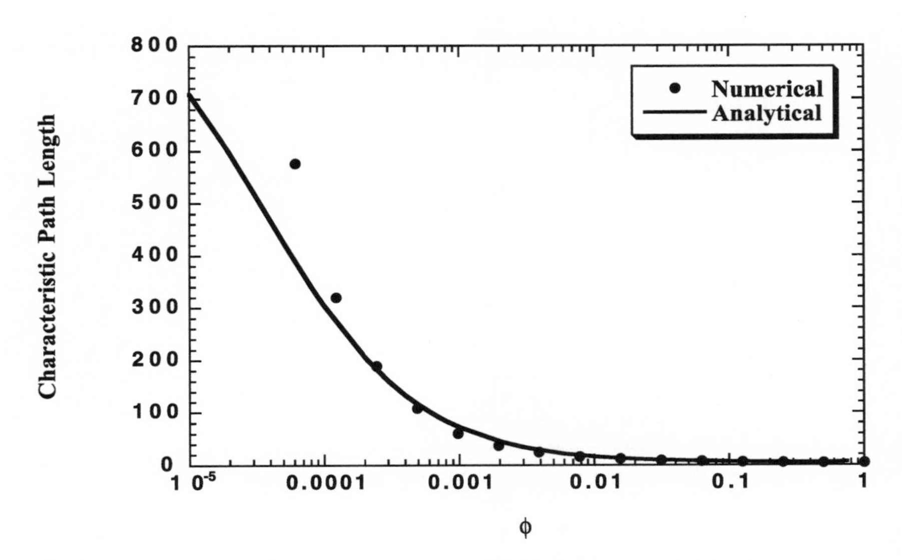

Figure 4.12 Comparison of analytical and numerical $L(\phi)$ for β-graphs ($n = 20{,}000$, $k = 10$).

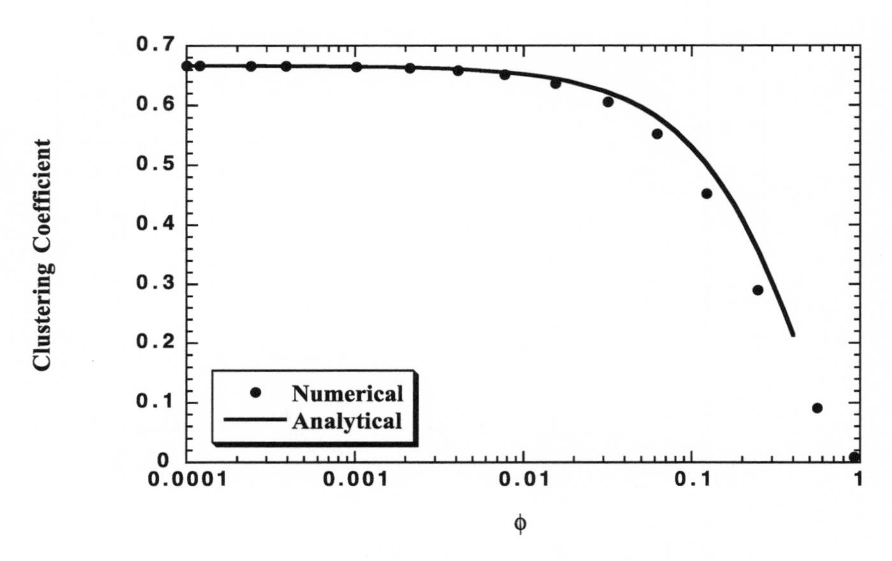

Figure 4.13 Comparison of analytical and numerical $\gamma(\phi)$ for β-graphs ($n = 1{,}000$, $k = 10$).

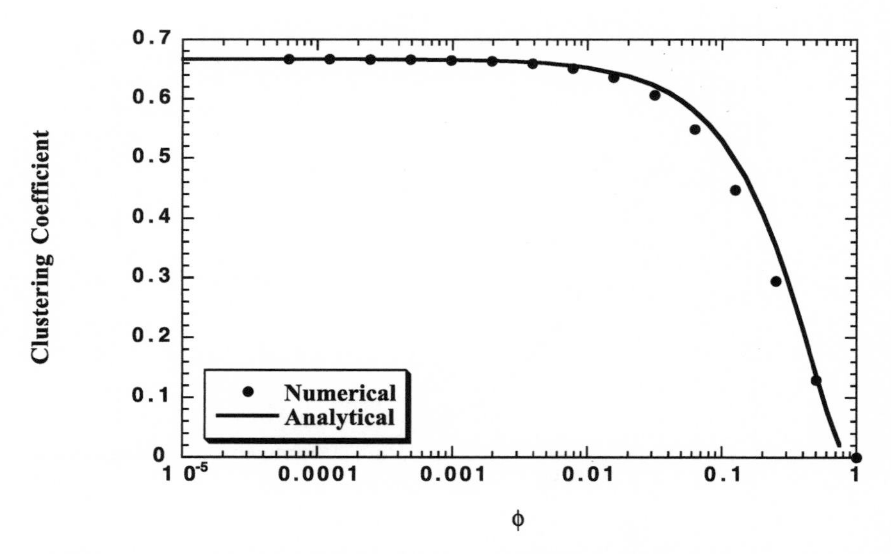

Figure 4.14 Comparison of analytical and numerical $\gamma(\phi)$ for β-graphs ($n = 20{,}000$, $k = 10$).

Finally, L and γ can also be approximated in terms of ψ, using Equation 4.17, where

$$\psi_{substrate} = \psi_{1-\text{lattice}} = \frac{2}{k(k-1)}.$$

Hence analytical approximations of $L(\psi)$ and $\gamma(\psi)$ can be compared with the numerical results, where the analytical version is computed parametrically as $L(\psi(\phi))$ and $\gamma(\psi(\phi))$, respectively (Figs. 4.15 and 4.16).

In all these cases, the analytical predictions compare favourably with their numerical counterparts, especially in the interval where ϕ is small (but not so small that the number of shortcuts in the β-model is unpredictable). Given also that Equation 4.15 exhibits the expected scaling phenomena and that Equations 4.15 and 4.16 possess the correct limiting values, it seems that the model presented here provides a plausible explanation of at least the broad-brush mechanisms driving the length and clustering statistics of the relational graphs considered in Chapter 3.

In fact, it is possible to go one step further and propose that the rapid transition of characteristic length from its "big-world" value near the 1-lattice to its "small-world" value is governed by the same mechanism that governs the length of a random graph as its average degree is increased. Recall from Section 4.1.2 that a pseudo-random graph exhibits a length contraction very similar to that of a Moore Graph with degree increasing from $k = 2$. An analogous mechanism appears to be at work in the relational-graph model: random edges added to an initial "ring," where the ring is the ring of meta-vertices at the global length scale. For small ϕ, what happens at the local length scale is trivial compared to the massive changes that occur with the addition of each global edge. Hence a graph that, locally, looks like a connected-caveman graph, a 1-lattice, or whatever else behaves globally just like a pseudo-random graph with $n = n_{\text{global}}$. When $k_{\text{global}} \approx \ln(n_{\text{global}})$, at which point the pseudo-random and random graphs are indistinguishable (Fig. 4.6), the relational graph has approached its random graph limit, and any subsequent changes in length must be small. This reasoning leads to one final prediction about relational graphs: that there is a critical value of ϕ (ϕ_{crit}) at which a relational graph has reached its random limit. If $k_{\text{global}} = \ln(n_{\text{global}})$ implies that the random limit has been reached, then, recalling that $k_{\text{global}} = 2(1-\phi) + k(k+1)\phi$, ϕ_{crit} can be

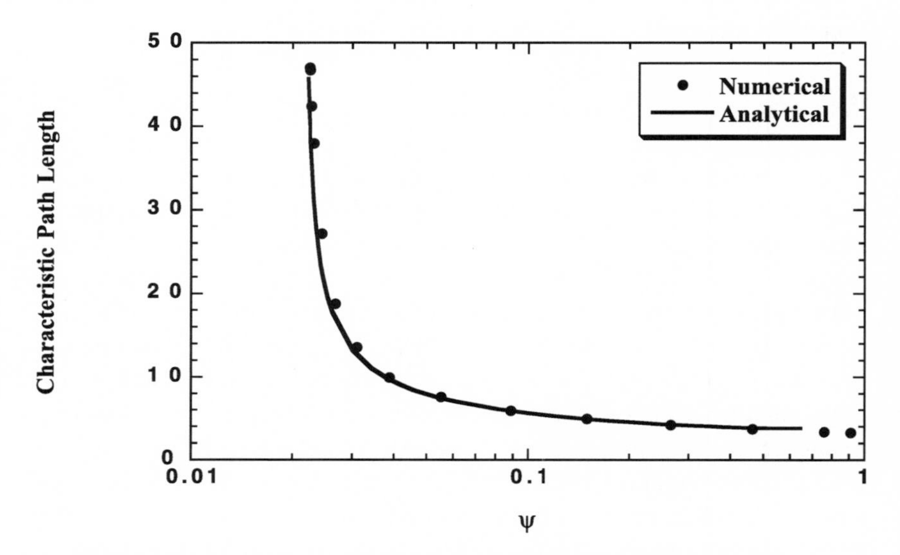

Figure 4.15 Comparison of analytical and numerical $L(\psi)$ for β-graphs ($n = 1{,}000$, $k = 10$).

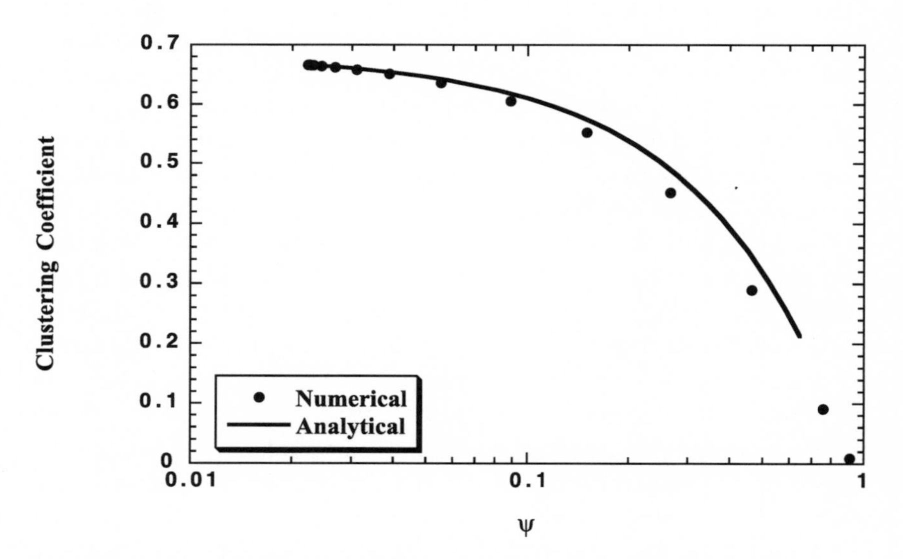

Figure 4.16 Comparison of analytical and numerical $\gamma(\psi)$ for β-graphs ($n = 1{,}000$, $k = 10$).

determined:

$$\begin{aligned} 2(1-\phi_{\text{crit}}) + k(k+1)\phi_{\text{crit}} &= \ln\left(\frac{n}{k+1}\right) \\ \Rightarrow \left[k(k+1)-2\right]\phi_{\text{crit}} &= \ln\left(\frac{n}{k+1}\right) - 2 \\ \Rightarrow \phi_{\text{crit}} &= \frac{\ln(\frac{n}{k+1}) - 2}{[k(k+1)-2]}. \end{aligned}$$

For the usual parameters of $n = 1{,}000$, $k = 10$, this yields $\phi_{\text{crit}} \approx 0.02$, and for $n = 20{,}000$ and $k = 20$, $\phi_{\text{crit}} \approx 0.05$, both of which seem reasonable estimates judging by Figures 4.11 and 4.12.

4.3 TRANSITIONS IN SPATIAL GRAPHS

We are now in a position to consider the models of spatial graphs that were introduced in the last part of Chapter 3. Is it possible that the models of the previous section can also be made to provide qualitative and quantitative agreement with the numerical results for spatial graphs and, in so doing, illuminate the qualitative differences between spatial and relational graphs? It turns out that both these goals can be attained without much adjustment to the models even as they stand now. First, however, the distinction between a length scale in a graph sense and a length scale in a physical sense must be clarified.

4.3.1 Spatial Length versus Graph Length

Up to this point, the word *length* has referred exclusively to the number of edges that must be traversed between two vertices in a graph. This terminology has been entirely deliberate as, in terms of structural properties of a general graph, this distance is the only one that has any unambiguous meaning. However, the introduction of spatial graphs in Chapter 3 raised the possibility that, for some graphs, the actual metric space in which the vertices exist could play a role in the construction of the graph and thus its structural properties. There is an important distinction to make here. *Graph distance* (represented by L for characteristic length and $d(i, j)$ for the shortest path length between two vertices) is still the only length measure that is considered a structural property of the graph. *Physical distance* determines (together with the parameter ξ) the probability of two vertices becoming connected during the

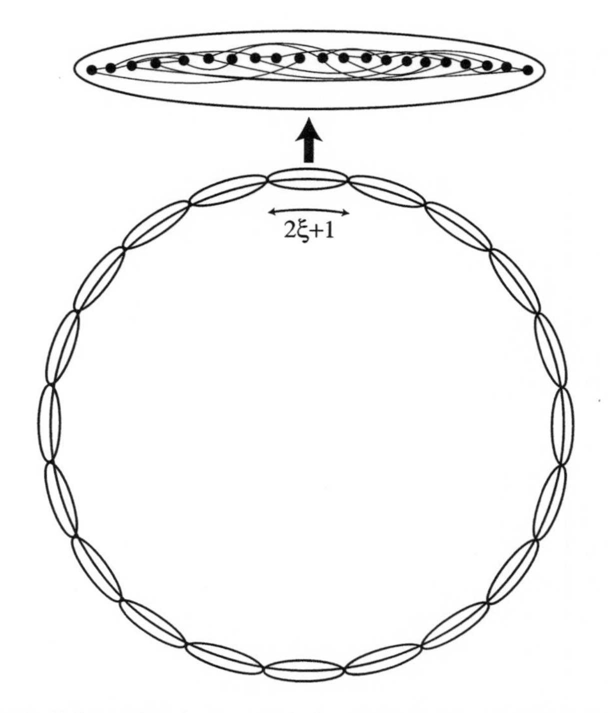

Figure 4.17 Schematic of characteristic path-length calculation for the uniform spatial-graph model.

construction process, much as the presence of preexisting edges determined the probability of connection in relational graphs. Hence physical distance may or may not (depending on ξ) closely correspond with graph distance.

4.3.2 Length and Length Scaling

Now the same method as before can be applied to the analysis of spatial graphs, where the vertices are assumed to be distributed uniformly in space and the local length scale is now defined by ξ instead of k. In one dimension, the required mental image is very similar to that used for a ring substrate above: imagine the 1-lattice to be divided into clusters containing ξ vertices, within each of which edges are formed with uniform random probability, and outside of which no connections can be made at all (see Fig. 4.17). Hence ξ determines the fraction of the world to which each vertex may connect and thus L_{local}.

This is the crucial distinction between relational and spatial graphs. In relational graphs, the number of vertices in a local neighbourhood remains fixed, and the local length scale is determined by the typical separation of those vertices. Hence there is a distinction between local edges (with $r = 2$), which connect vertices within the same local neighbourhood, and global edges, which connect vertices in different neighbourhoods. In spatial graphs, however, the number of vertices included in the same local neighbourhood is determined by ξ, where ξ is measured according to some *external* metric. Because vertices can connect only to other vertices that are local, according to this external metric, then *all edges are local*, regardless of what their range is. This property of spatial graphs reflects the fact that edges with $r > 2$ do not necessarily connect vertices that are separated on a global length scale. Rather, they connect vertices chosen at random from within the *external length scale*, parameterised by ξ.

In spatial graphs, therefore, the overlap of neighbourhoods (which, was ignored in the relational-graph model) becomes increasingly significant as ξ increases. One way to compensate for neighbourhood overlap is to reformulate the expression for L_r in such a way that, when estimating the distance between vertices in different neighbourhoods, only the intermediate neighbourhoods are counted. Hence,

$$L_s = L_{\text{local}} + (L_{\text{global}} - 1)(L_{\text{local}} + 1), \tag{4.19}$$

where now

$$\begin{aligned} k_{\text{local}} &= k, \\ n_{\text{local}} &= 2\xi + 1, \\ k_{\text{global}} &= 0, \\ n_{\text{global}} &= \frac{n}{2\xi + 1}. \end{aligned} \tag{4.20}$$

Because $k_{\text{global}} = 0$, Equation 4.19 can be simplified using Equation 4.10 to obtain the equivalent expression of Equation 4.15 for spatial graphs:

$$\begin{aligned} L_s &= L_{\text{M}}(n_{\text{local}}, k_{\text{local}}) + \left(\frac{n_{\text{global}}^2}{4(n_{\text{global}} - 1)} - 1\right)[L_{\text{M}}(n_{\text{local}}, k_{\text{local}}) + 1] \\ &= L_{\text{M}}\left(2\xi + 1, k\left(1 - \frac{2}{k+1}\right)\right) + \left(\frac{n^2}{4(2\xi+1)(n - 2\xi - 1)} - 1\right) \\ &\quad \times \left[L_{\text{M}}\left(2\xi + 1, k\left(1 - \frac{2}{k+1}\right)\right) + 1\right]. \end{aligned} \tag{4.21}$$

Note that, in contrast to relational graphs, the driving parameter controls not the number of edges in the local and global scales, but the relative sizes of the two scales. In other words (to restate the main point), for relational graphs the transition from structure to randomness occurs through the conversion of *local edges* to *global edges*. In spatial graphs, nothing of the sort happens. Rather, all edges remain local at all times, except that the local scale grows until it becomes indistinguishable from the global scale. This is why shortcuts and contractions are unable to reconcile the differences between spatial and relational graphs: there was an implicit assumption in the definition of both quantities that if connections were made outside the local scale, then they existed on the global scale and hence could contract distances on the order of the length of the entire graph. This is what caused graphs with only a small ϕ or ψ to have lengths comparable to those of equivalent random graphs. However, in spatial graphs shortcuts or contractions tend only to contract pairs of vertices that are separated by distances just slightly greater than the cutoff of $d(i, j) = 2$, so the resulting length contractions are small. Equation 4.21 also explains why spatial graphs show linear length scaling with respect to n. For any fixed ξ, increasing n causes n_{global} to increase linearly, and, as k_{global} remains fixed at $k_{\text{global}} = 0$, L_{global} also increases linearly according to Equation 4.21. Hence, although L_{local} scales logarithmically with respect to n, L is still dominated by the linear L_{global} term.

4.3.3 Clustering

The key idea in computing the clustering coefficient for spatial graphs, using this model, is that k_{local} stays fixed even as ξ (and consequently n_{local}) increases. Hence the same number of edges gets spread uniformly over an increasing number of vertices in the local scale. This again is quite different from relational graphs, where n_{local} stays fixed and k_{local} decreases with increasing ϕ. The clustering coefficient (γ_s) can be computed as follows.

Each vertex v has k neighbours ($\Gamma(v)$) within its *spatial neighbourhood* $\Gamma_\xi(v)$, which contains all vertices $v - \xi \leq u_i \leq v + \xi$. Hence the probability that any given u_i is also a neighbour of v is $k/2\xi$. If $u_i \in \Gamma(v)$, then it matters how many other $w \in \Gamma(v)$ are adjacent with u_i, as this is the number of edges that each u_i contributes to $\gamma(v)$. The *overlap of neighbourhoods* $\mu(v, u_i)$ can then be defined as $\mu = |\Gamma(v) \cup \Gamma(u_i)|$, where the probability of any vertex $w \in \Gamma(u_i)$ also being an element of $\Gamma(v)$

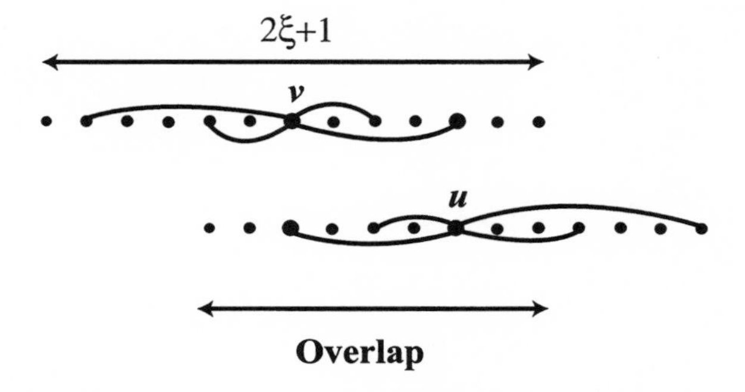

Figure 4.18 Calculation of the clustering coefficient (γ) for the uniform spatial-graph model involves the overlap of the two neighbourhoods $\Gamma(u)$ and $\Gamma(v)$ defined by ξ.

is $(k/2\xi)^2$, as the distribution of edges in $\Gamma(v)$ and $\Gamma(u_i)$ is independent and uniform (see Fig. 4.18). Then the expected number of edges contributing to $\gamma(v)$ from u_i is

$$\text{Clustering edges from } u_i = P\big[u_i \in \Gamma(v)\big] \cdot \mu(v, u_i) \cdot P\big[w \in \Gamma(v) \cup \Gamma(u_i)\big]$$
$$= \frac{k}{2\xi} \cdot \mu(v, w_i) \cdot \left(\frac{k}{2\xi}\right)^2.$$

Summing over all possible u_i, where $(\xi-1) \le \mu \le 2(\xi-1)$, the expected number of clustering edges in $\Gamma(v)$ is[4]

$$\text{Clustering edges in } \Gamma(v) = \left(\frac{k}{2\xi}\right)^3 \sum_{\mu=\xi-1}^{2(\xi-1)} \mu$$
$$= \frac{3}{2}\left(\frac{k}{2\xi}\right)^3 \xi(\xi-1)$$
$$= \frac{3}{16}\left(\frac{k}{\xi}\right)^3 \xi(\xi-1).$$

Dividing by the total number of possible edges in $\Gamma(v)$ yields the clustering coefficient:

$$\gamma_s = \frac{2}{k(k-1)} \frac{3}{16}\left(\frac{k}{\xi}\right)^3 \xi(\xi-1)$$
$$= \frac{3}{8}\frac{k^2(\xi-1)}{\xi^2(k-1)}. \tag{4.22}$$

As expected, (for $k \gg 1$, $\xi \gg 1$), $\gamma_s \propto k/\xi$. Also, for $\xi = O(n)$, $\gamma_s = O(k/n)$ as expected in the random-graph limit. Furthermore, for $\xi = k/2$

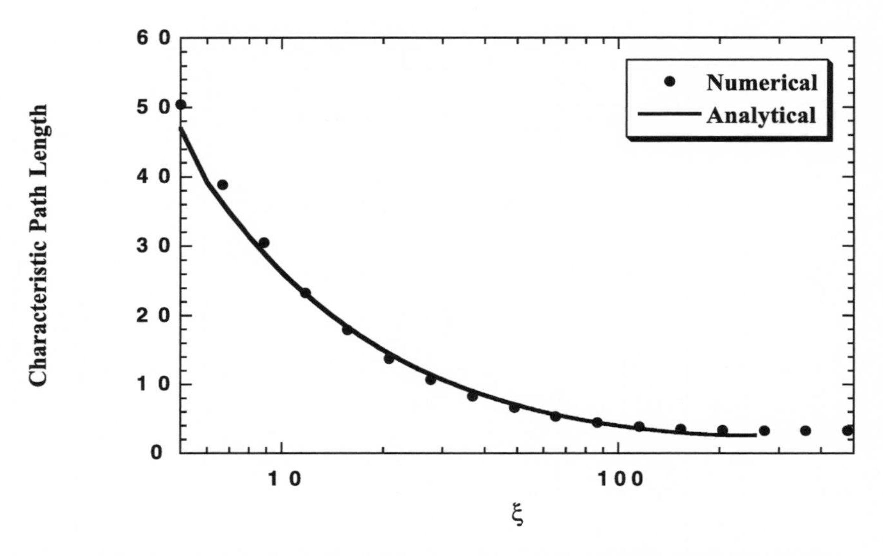

Figure 4.19 Comparison of analytical and numerical $L(\xi)$ for uniform spatial graphs with $n = 1{,}000$, $k = 10$.

(the lower limit of ξ, equivalent to a 1-lattice), Equation 4.22 reduces to

$$\begin{aligned}\gamma_s\big|_{\xi=\frac{k}{2}} &= \frac{3}{8}\frac{k^2(\frac{k}{2}-1)}{\frac{k^2}{4}(k-1)}\\ &= \frac{3}{4}\frac{(k-2)}{(k-1)}\\ &= \gamma_{1\text{-lattice}}.\end{aligned}$$

4.3.4 Results and Comparisons

The previous section demonstrates that both L_s and γ_s have the correct limiting values and scaling properties, but it remains to be confirmed that the expressions exhibit quantitative agreement with the numerical data of Chapter 3 for the entire range of ξ. Figures 4.19 and 4.20 show the relevant comparisons between predicted and actual L and γ for the uniform spatial-graph model. As with the relational-graph approximations, the agreement is close and certainly has the correct qualitative features. The most important of these is that $L(\xi)$ and $\gamma(\xi)$ have the *same shape* as each other, a characteristic that *precludes the possibility of small-world spatial graphs*. Hence the intuitive argument of Chapter 3 appears to be confirmed by these analytical constructions.

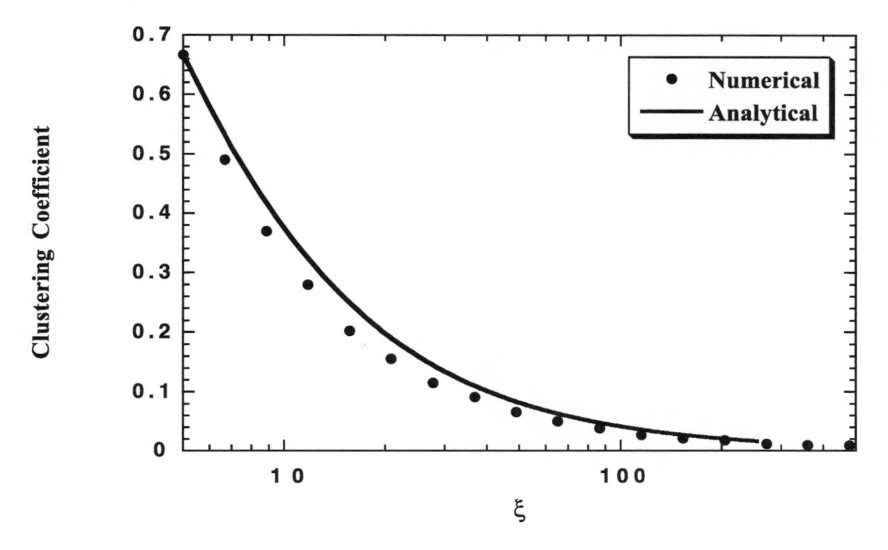

Figure 4.20 Comparison of analytical and numerical $\gamma(\xi)$ for spatial graphs with $n = 1{,}000$, $k = 10$.

4.4 VARIATIONS ON SPATIAL AND RELATIONAL GRAPHS

It is now possible to put forward a tentative model of what might be called *transitional graphs*: graphs that span the range of clustering and length properties between highly clustered, ordered and highly expanding, random limits but do so in a manner that combines both the spatial and relational mechanisms considered above. This combination seems appropriate because "real" graphs (that is, graphs that map the connectivity of real, observable networks) are likely, in general, to have some elements of both mechanisms. This will become apparent in the next chapter, but even now it seems likely that a network that owes its existence to both human and physical (such as geographical) influences might well exhibit both relational and spatial characteristics. That is, there may well be a characteristic spatial-length scale over which most connections are likely to occur, but occasional edges may be formed, either at random or as a function of preexisting edges, at length scales on the order of the entire graph. Hence both mechanisms may be necessary in general. In fact, real graphs are probably vastly more complicated than even this more general approach, exhibiting multiple length scales or even hitherto unexplored mechanisms. Nevertheless, the following formulation of *transitional graphs* may be a start:

$$L_t = L_{\mathrm{M}}\left(n_{\text{local}}(\xi), k_{\text{local}}(\phi_g)\right) + L_{\mathrm{M}}\left(n_{\text{global}}(\xi), k_{\text{global}}(\phi_g)\right)\left[L_{\mathrm{M}}\left(n_{\text{local}}(\xi), k_{\text{local}}(\phi_g)\right) + 1\right] \tag{4.24}$$

and

$$\gamma_t = \frac{k_{\text{local}}\left(k_{\text{cluster}} - 1\right)}{k(k-1)}, \tag{4.25}$$

where

$$\begin{aligned} k_{\text{local}} &= \left(1 - \phi_g\right)k, \\ n_{\text{local}} &= 2\xi + 1, \\ k_{\text{global}} &= (2\xi + 1)k\phi_g, \\ n_{\text{global}} &= \frac{n}{2\xi + 1}, \\ k_{\text{cluster}} &= \left(1 - \phi_g\right)\left[\frac{3}{8}\frac{k^2}{\xi^2}(\xi - 1) + 1\right], \end{aligned} \tag{4.26}$$

and where ϕ_g is the fraction of *global edges* in the graph. The intuition behind a global edge is clear: an edge that connects vertices that would otherwise be separated by a distance on the order of the length of the graph (that is, $d\,(i, j) = O\,(L\,(G))$). This new definition is required because, as we recall, the previous notion of shortcut edge proved insufficiently general to deal with spatial graphs, which required many more shortcuts than relational graphs before they exhibited small characteristic path lengths. The reason for this was that the shortcuts in spatial graphs had small range because they were connecting only vertices in the same *spatial neighbourhood*. An ideal definition would distinguish between shortcuts with small range and shortcuts with a range of $O\,(L\,(G))$. Unfortunately, this is a somewhat slippery criterion, as all the length scales involved (local and global) are continually changing as a result of the shortcuts themselves. Nevertheless, the following definition, although inadequate, may lead to a better formulation in the future.

Definition 4.4.1. The spatial neighbourhood of a vertex v $\left(\Gamma_\xi\,(v)\right)$is the set of vertices within radius ξ of v in d-space.

Definition 4.4.2. The *diameter* of $\Gamma_\xi\,(v)$ $\left(D\left(\Gamma_\xi\,(v)\right)\right)$ is the largest expected path length between any two vertices in the same spatial neighbourhood.

Definition 4.4.3. A *global edge* is an edge (u, v) with range $r(u, v) > D(\Gamma_{\xi}(v))$.

That is, a global edge is an edge that connects vertices that could not have been in the same spatial neighbourhood; hence they must be separated by a distance on the order of the entire graph. Thus, for small ξ, spatial graphs will exhibit shortcuts (as previously defined) but no global edges, whereas almost all shortcuts in relational graphs will also be global edges. The main problem with this definition is that, for large ξ or ϕ_g, the local and global length scales converge, at which point *all edges* will have $r = O(L(G))$, whilst simultaneously *no edges* will have $r > D(\Gamma_{\xi}(v))$, so all edges will be both local *and* global. This is a quandary that has not been resolved, so no results will be presented on this matter.

Another variation upon the theme of spatial versus relational graphs arises if one admits probability distributions with *infinite variance*: that is, if a graph were constructed according to the spatial-graph recipe but using, say, a Cauchy distribution instead of a uniform or Gaussian distribution. In terms of the parameter ξ, the Cauchy distribution would result in the probability of connection

$$P(i, j) = \frac{\xi}{\pi(\xi^2 + x_{i,j}^2)}.$$

Note that the tails of this distribution decay only algebraically, not exponentially as is the case for a Gaussian. This slow-decay feature is what causes the infinite variance and (as with the transitional-graph formulation above) raises the possibility that, despite the existence of an externally imposed length scale (ξ) over which *most* connections occur, *some* connections will always have a range on the order of $L(G)$. One might expect then that such a distribution would generate small-world graphs, and Figure 4.21 indicates that this is precisely what happens. There is a significant difference, however, between these *Cauchy Graphs* and all the models considered previously: it is impossible to construct Cauchy Graphs with no shortcuts. Hence the 1-lattice limit (or even an approximation thereof) is unattainable, and *every* Cauchy Graph is "small" regardless of ξ. In other words, the picture presented by Figure 4.21 is similar to Figure 3.22 with the left-hand end chopped off. For this reason, in Figure 4.21 $L(\xi)$ and $\gamma(\xi)$ are scaled by their random-limit values ($\xi = n/2$) instead of their values at the 1-lattice limit. Nevertheless, it is clear that for a broad interval of ξ, the Cauchy distribution yields small-world graphs. This observation suggests that all that is required of a small-world graph is that, first, vertices have a better than random

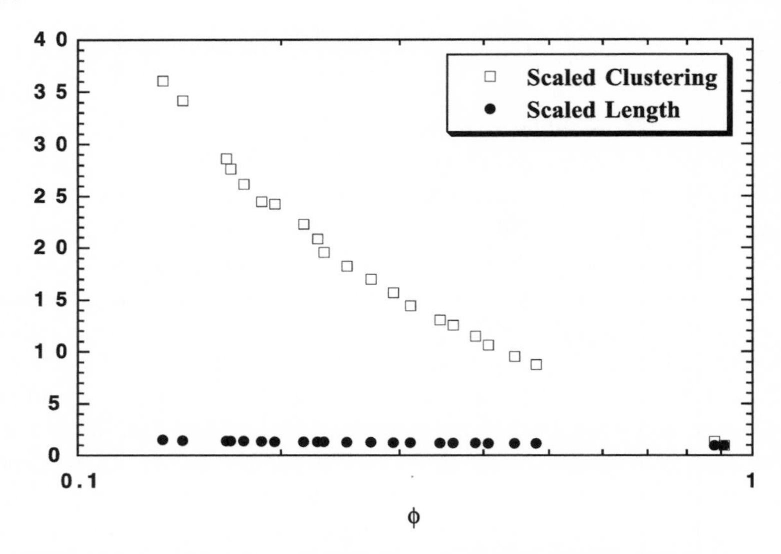

Figure 4.21 $L(\xi)$ and $\gamma(\xi)$ for spatial graph with a Cauchy distribution ($n = 1{,}000$, $k = 10$). L and γ are scaled by their random-limit values.

chance of forming triads. Whether this preference results from spatial proximity (as with the Cauchy Graphs) or pre-exisiting relationships (as with the α- and β-graphs) appears to be immaterial. But, second, despite this, some edges have a range that is $O(n)$.

If this is true, the small-world phenomenon may turn out to be very general indeed, applying not just to relational graphs and Cauchy Graphs in one dimension, but perhaps to a broad range of spatial graphs, in arbitrary dimension d, with distributions of the form

$$P(i, j) = \frac{\xi}{\pi(\xi^a + x_{i,j}^a)}.$$

A general and precise characterisation of small-world graphs, along with the relationship between a, d, and the small-world phenomenon, appear to be interesting open research questions.

4.5 MAIN POINTS IN REVIEW

1. *Small-world graphs* are redefined in a model-independent fashion, which requires no knowledge of the construction of the graph or, in fact, the construction of any graphs. All that is required in order to determine whether or not a graph is a small-world graph is

its corresponding n and k. This is a stronger result than that obtained in Chapter 3, in which a construction algorithm was implicitly required.

2. Based on the characteristics of caveman Graphs and Moore Graphs, an analytical model is developed, which predicts the following features of spatial and relational graphs, as observed numerically in Chapter 3:

    a. Limiting values of L and γ at ordered and random extremes.
    b. Functional forms of L and γ between these limits.
    c. Length-scaling properties with respect to n and k as a function of ϕ and ξ, respectively.

3. As a result, the different qualitative features of relational and spatial graphs can be explained, including why only relational graphs can be small-world graphs. This conclusion, however, is predicated on the assumption that the defining probability distribution for a spatial graph has, effectively, a finite cutoff. Distributions that do not have this property (such as the Cauchy distribution, with infinite variance) apparently can exhibit the small-world phenomenon, but the general case remains an open problem.

5

"It's a Small World after All": Three Real Graphs

Kevin Bacon is not, as it turns out, the first person to have his name associated with the small-world phenomenon. Appropriately this distinction belongs to the great twentieth-century mathematician and co-founder of probabilistic graph theory Paul Erdös. Erdös was a prolific writer as well as an astonishing mathematician and famously eccentric personality, authoring or co-authoring over fourteen hundred papers in his lifetime (and even some after). Naturally enough, a person's *Erdös Number* ($E(v)$) is a measure of how close they are to Erdös via their co-authors. An Erdös Number of one means that you have published with Erdös directly (472 people can boast this), an Erdös Number of two implies that you haven't published a paper with him but that you have published with someone else who has, and so on.

Following this reasoning, one can construct what is now known as the *collaboration graph*, in which the vertices are publishing academics (in any field at all) and the edges represent co-authorship. This fascinating construction has been the subject of some attention from graph theorists recently,[1] but, because of its daunting size, only a tiny fraction of it has ever even been mapped out, let alone analysed. Not surprisingly, this fraction is always chosen to be in Erdös's local vicinity (see Grossman and Ion 1995 for details). Using data available on the World-Wide Web one can calculate $\gamma_{\text{erdös}}$.[2] There are 492 authors in Erdös's neighbourhood, and, on average, each is connected with 5.76 other members of $\Gamma_{\text{erdös}}$. Hence $\gamma_{\text{erdös}} = 5.76 \cdot 492/492(492 - 1) \approx 0.012$. This may not seem all that high, but we should bear the following facts in mind. First, Erdös has many more co-authors ($k_{\text{erdös}}$) than nearly anybody else in the history of mathematics, and so his neighbourhood is almost certainly more sparsely connected than average. Second, this is still much greater than we would find for a random graph, where $\gamma \approx k/n$. Even if everyone had $k = k_{\text{erdös}}$ and even if there were, say, only $n = 100{,}000$ publishing authors in the history of all the sciences (which is probably a significant underestimate), then $\gamma_{\text{random}} = 0.00492$, which is already much less than $\gamma_{\text{erdös}}$.

Given this (admittedly highly unrepresentative) example, we might conclude that the collaboration graph is highly clustered. It is also thought not only that the Erdös Component (the graph of all authors who have a finite Erdös Number) encompasses most of the mathematical, physical, and social sciences, but also that its characteristic path length is quite small (Grossman and Ion 1995). Hence we might speculate that world of Paul Erdös is indeed a small one.

We would, however, just be speculating. And given the truly vast amount of data collection that would be required to make any comprehensive assessment of the entire collaboration graph, such speculations are likely to remain just that for some time. What would be ideal then is a graph (or graphs) that, like the collaboration graph, represent well-defined connections of real networks and that are of either social or scientific interest, but that are completely known and mapped out. Such a request may seem a difficult one to fulfill, especially given that any such graph would also have to be fairly large and sparse (but still connected) in order to say anything interesting about it. For instance, a typical social network would be large and sparse enough, but the connections of social networks are notoriously difficult to pin down, and their extent is often very hard to define (imagine, for instance, trying to map out the friendship network of New York City). At the other extreme, a graph representation exists of the different regions of the visual cortex of a macaque monkey (Felleman and Van Essen 1991), in which the edges are both well defined and known. But for this example $n = 32$ and $k = 12$; hence any connected topology would result in almost the same characteristic length and clustering, so not much can be said.

Nevertheless, we *can* find three quite diverse graphs, both in terms of their parameters (n, k) and in terms of the nature of the networks they represent, which do satisfy the requirements and which exhibit some interesting properties:

1. The “Kevin Bacon Graph,” which is the Hollywood equivalent of the collaboration graph, in which a connection implies that two actors have acted in a movie together.
2. The “Western States Power Graph,” which, as the name suggests, maps the power stations and high-voltage transmission lines that supply power to all the states west of the Rocky Mountains.
3. The “*C. elegans* Graph,” which represents the neural connections of the celebrated and much-studied worm *Caenorhabditis elegans*.

In each of these cases, a complete adjacency list of the relevant graph is available, from which all statistical properties can be measured directly. In this chapter each system under consideration is discussed (with apologies to all experts in those respective fields), its structural properties are analysed using the methods developed in Chapter 3, and then the results are compared with the predictions of the theoretical models of Chapter 4, from which a characterisation of each graph can be made. The general question at issue here is twofold: Do small-world graphs actually exist? and Can the models developed in Chapters 3 and 4 adequately characterise the statistics of real graphs? In other words, are the phenomena and mechanisms of the previous chapters merely artifacts of theoretical constructions, or can they help us predict and understand phenomena that are actually observable in the real world?

5.1 MAKING BACON

To date, the history of motion pictures has resulted in the creation of roughly 150,000 films, boasting a combined cast of over 300,000 actors[3] from every movie-making country on earth, dating back to the dawn of motion pictures,[4] even before the introduction of sound. All this information is located in a single, searchable database, the Internet Movie Database (IMDb), which is readily accessible (at www.us.imdb.com) to anyone with a web browser and plenty of spare time. It is easy to see how one might create a graph from this information: every actor is a vertex, and every joint appearance of two actors in the same film is an edge (multiple joint appearances are treated as a single edge). The result is what is called here the *Kevin Bacon Graph* (KBG).[5] This may sound like somewhat whimsical (or even bizarre) subject matter for the serious business of mathematics, but consider the following:

1. The KBG is well defined in terms of its extent, in that it contains only actors in feature films, all of which are known. So there is no ambiguity about who is and isn't included.
2. All edges in the graph are also both well defined (either two actors were in a movie together or they weren't) and known.
3. About 90% of all actors in the database are included in a single *connected component* consisting (as of April 1997) of about 225,000 actors in about 110,000 films. It then suffices to treat this component as the Kevin Bacon Graph, bearing in mind that only connected graphs are being considered here.

4. The connected component is sufficiently large ($n = 225,226$) and sparse ($k \approx 61$) that, depending on its structural characteristics, it could exhibit a characteristic length spanning several orders of magnitude. Likewise, its clustering coefficient could also potentially exhibit any value between almost 1 and almost 0. Hence it offers a clear test of the models from Chapters 3 and 4 that could be, on the one hand, approximately correct or, on the other, hopelessly wrong.
5. It is not *so* large that it cannot be stored or manipulated by a (powerful) computer. Specifically, if n were even a single order of magnitude larger, then it would be difficult to find *any* computer that could hold the whole thing in memory at once,[6] a necessary feat if one wishes to compute, say, characteristic length in any reasonable amount of time.

Hence it turns out that, far from being silly, the KBG is actually close to an ideal graph for the purposes outlined above. Furthermore, it is almost identical in nature to the more serious, but far less accessible, collaboration graph in which "collaboration" has simply been redefined to mean "acting in a movie together" rather than "publishing a paper together."[7] Thus justified, the relevant statistics of the KBG can now be computed.

5.1.1 Examining the Graph

Before computing anything, however, there is one structural feature of the graph that upon reflection, is obvious. Since the criterion for an edge is that two actors have acted together in a movie, then each movie represents a complete subgraph in which every cast member of that film is adjacent with every other cast member. *This property of the graph has the important consequence that it almost completely eliminates the possibility of shortcuts.* In fact, the only possible situation in which a shortcut can occur is if there is a movie with a cast of only two people. To be sure, in the dark recesses of movie history, it seems possible that this has indeed happened, but we cannot reasonably expect such aberrations to be responsible for any small-world properties. Fortunately this eventuality was anticipated in the earlier discussion of *contractions*, which turn out to be precisely the construction required here: individual actors bring otherwise disparate *groups of actors* close together via their collaboration on individual films. All actors who have worked on multiple

TABLE 5.1
Statistics computed for the KBG

	Value
n	225,226
k	61
L	3.65
γ	0.79 ± 0.02
ϕ	$0.0002 \pm 6 \times 10^{-6}$
ψ	0.166 ± 0.005

films cause contractions to some extent (recall that contractions occur even in very regular "big-world" topologies), but a minority of actors are instrumental in this sense. These actors might be called *linchpins*. An example of one type of linchpin—a *temporal linchpin*—is Eddie Albert, of *Green Acres* fame, who has appeared in over eighty films, spanning a sixty-year career. He links together such greats as Humphrey Bogart, Marlon Brando, Richard Burton, John Travolta, and (of course) Kevin Bacon.[8] Another kind of linchpin—a *cultural linchpin*—is one that spans movies made in different countries. An obvious example is Bruce Lee, who links the Chinese acting industry to Hollywood (and hence the rest of the world) through classics like *Game of Death* and *Enter the Dragon*, both of which also starred Chuck Norris. Finally, there are *genre linchpins* who span different kinds of films: comedy and suspense (Nicholas Cage in *Raising Arizona* and *Kiss of Death*), action and drama (Mel Gibson in the *Lethal Weapon* and *Hamlet*), and even historical and tech-noir (Sigourney Weaver in *Gorillas in the Mist* and the *Alien* series). Despite these occurrences, most actors act mostly with their co-stars' co-stars, leading to a highly locally clustered network in which a small fraction of individuals do most of the work of bringing everybody else close together.

But let us see this for certain. Running the same routines used in Chapter 3 on imaginary graphs, the statistics shown in Table 5.1 can be computed for the KBG. To obtain these, L was computed exactly,[9] and γ, ϕ, and ψ were approximated by taking a random sample of a thousand, vertices and calculating the appropriate statistics in their neigbourhood.

As expected, γ is high, ϕ is negligible, and ψ is small but significant. Also, most importantly, L is small: small, that is, in comparison to the length of the equivalent caveman graph ($L_{\text{caveman}} \approx 1{,}800$) and almost as small as an equivalent random graph ($L_{\text{random}} \approx 3$), as approximated by a Moore Graph. Hence it is probably safe to conclude that the KBG is a *small-world graph* in that it has approximately the same length as an equivalent random graph but is orders of magnitude more clustered (a random graph would have $\gamma_{random} \approx k/n = 0.00027$).

5.1.2 Comparisons

Having satisfied the broad claim that small-world graphs actually exist (or at least one does), the next issue is to determine whether or not the *mechanisms* proposed in Chapter 4 can explain these properties; that is, do the data fit the theoretical curve? Before answering this, it is worth pointing out how unlikely such a comparison is to succeed, given the assumptions implicit in the models. Perhaps the one that is most obviously incorrect is the assumption that all vertices have the same degree k: in the KBG, the degree ranges $1 \leq k \lesssim 3000$. Nevertheless, let us make the following comparison: for the parameters $n = 225{,}226$ and $k = 61$, generate a single curve for $L(\psi)$ (and another for $\gamma(\psi)$) using the model of relational graphs presented in Chapter 4;[10] on the same figure, place the two points, $L(\psi_{\mathrm{KBG}})$ and $\gamma(\psi_{\mathrm{KBG}})$, from Table 5.1. Note that there is only one free parameter in all this: b, the bundle size, which is roughly equivalent to the average number of cast members in a movie. This number is unknown for the real graph (although an average value could be calculated with some effort), but fortunately the model is not very sensitive to variations of as much as an order of magnitude in b, so the precise value does not matter much ($b = 10$ is used here). In contrast, however, it is possible for L and γ to fall anywhere between $1{,}800 \gtrsim L \gtrsim 3$ and $1 \gtrsim \gamma \gtrsim 0.00027$ and ψ to be anywhere in the range $0 \leq \psi \leq 1$. Effectively then, this comparison contains *no free parameters*, so the fact that both $L(\psi_{\mathrm{KBG}})$ and $\gamma(\psi_{\mathrm{KBG}})$ fall as close to their respective curves as they do is quite remarkable (see Figs. 5.1 and 5.2).

What is more remarkable, though, is that *no other plausible explanation that has been considered* can come anywhere near as close to the real data points as the predictions of the relational-graph model. As Table 5.2 shows, both the caveman-graph model and a random-graph approximation fail dreadfully on one of the statistics (L or γ) and do worse than the relational-graph model on the other.

A final comparison to make is with the other class of model under consideration: spatial graphs. One might not expect this model to compare as well to the data as the relational-graph model, as it is premised on systems that exhibit a characteristic spatial scale at which connections are likely to be made. There is no plausible reason to think that this is true for the KBG, but the only way to check this is to show that the

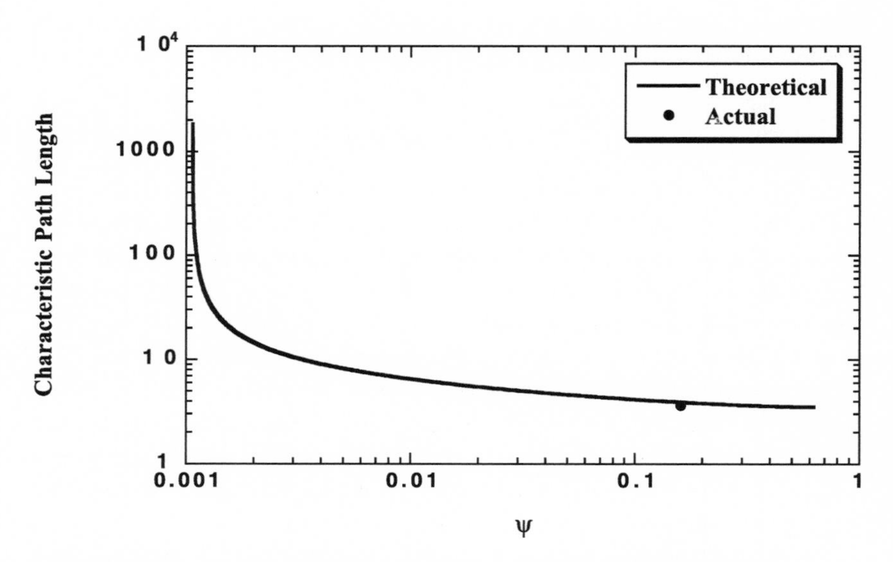

Figure 5.1 Theoretical prediction for $L(\psi)$ vs. ψ for KBG parameters ($n = 225{,}226$ and $k = 61$) with actual $L(\psi_{\text{KBG}})$ superposed.

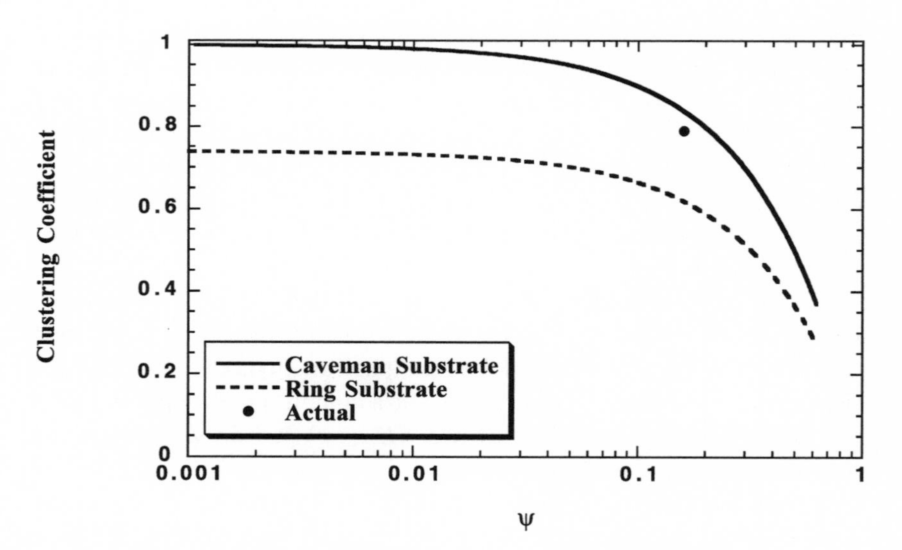

Figure 5.2 Theoretical prediction for $\gamma(\psi)$ vs. ψ for KBG parameters ($n = 225{,}226$ and $k = 61$) with actual $\gamma(\psi_{\text{KBG}})$ superposed. Predictions based on both connected-caveman and ring substrates are shown for comparison.

TABLE 5.2
Comparison of Statistics Computed for the KBG with Those Predicted by the Connected-Caveman and Random Graph Models

	KBG	Relational ($\psi = 0.166$)	Caveman	Random
L	3.65	3.9	1,817	2.99
γ	0.79	0.61–0.84	0.999	0.00027

model does not produce reasonable predictions for *any* ξ. Note that this is a much weaker test than that posed for the relational-graph model, where ψ could actually be measured from the graph itself. Because the graph includes no explicit spatial information, any characteristic spatial scale would have to be expressed implicitly by the edge set, so the model cannot be tested for a specific value of ξ. For this reason, it would be hard to place much confidence in a positive result: even if the model did fit for some value of ξ, that would not necessarily mean that the model was a good explanation of the phenomenon. However, a *negative* result *is* possible—if it turns out that the L and γ values cannot be satisfied for *any* ξ, then the spatial model definitely fails.

Figure 5.3 shows that this negative result is indeed the case for a one-dimensional, uniform spatial model. Here L is plotted as a function of γ, where both are parameterised by ξ. The actual (γ_{KBG}, L_{KBG}) value is superposed and clearly does not lie near the curve for any value of ξ. Furthermore, as an increase in the dimension of a spatial model generally results in lower γ, the comparison does not improve simply by raising the dimensionality. Hence the only model that produces a satisfactory fit to the data is the relational-graph model, a fact that may or may not seem surprising, depending on your intuition. Clearly the model was designed to simulate just this kind of system, so, on the one hand, it is not so surprising that it does just that. On the other hand, some rather drastic assumptions were made along the way in order to derive a tractable model, so it *is* surprising that it should match up so well with *any* real system's statistics.

By the way, for those who are interested, Kevin Bacon is *not* the centre of the Hollywood Universe, despite all claims to the contrary—Rod Steiger is. In fact, although he is one of the better-connected actors in the database, Kevin Bacon turns out to be nothing special (a mere 669th on the list of the all-time greats).[11] This is something of a double irony,

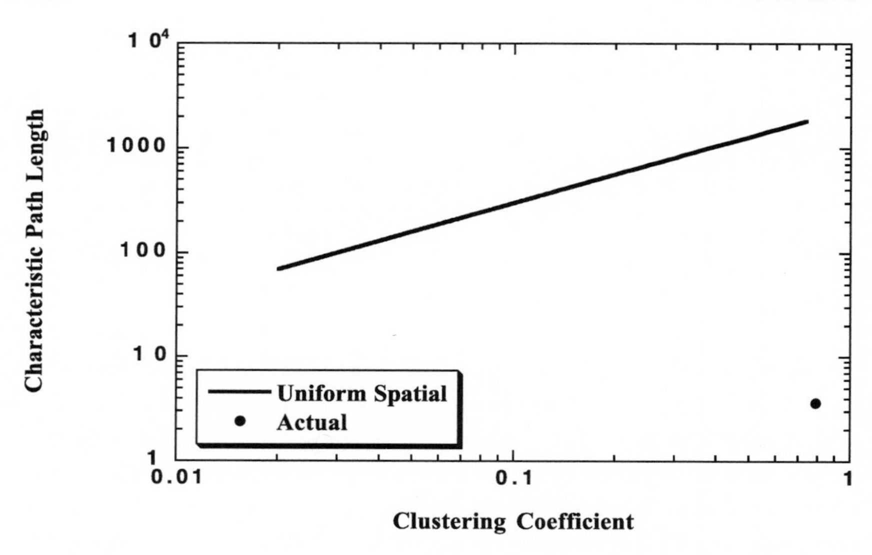

Figure 5.3 Plot of $L(\xi)$ vs. $\gamma(\xi)$, parameterized by ξ. No value of ξ satisfies both computed statistics of the KBG (L_{KBG}, γ_{KBG}).

as the novelty of the Kevin Bacon Game itself originally derived from Kevin Bacon's image as being nothing special: hence the surprise that he was some how at the centre of everything. But in a way, Bacon's ambiguous status is also the main point: in a small world, *everyone* seems to be at the centre, because everyone is close to everyone else. One way to see this is to plot the distribution sequence for Kevin Bacon. Recall that this is just the number of actors who can be reached from Bacon at each degree of separation (a.k.a "Bacon Number"), where acting together in a movie counts as one degree.

Figure 5.4 shows Bacon's distribution sequence compared with that of the *real* centre (Steiger) and also the *average* distribution sequence over all actors in the database. The most striking feature of this histogram is that all the distributions look roughly the same. Steiger is clearly better than Bacon, who is clearly better than average, but in all three cases the bulk of the graph has been reached within four degrees, and in no case is it reached in less than three. Compared with the path lengths of, say, the connected-caveman graph, a difference of one degree of separation is trivial, and it would have been no less surprising to movie buffs if almost any actor could be connected to Kevin Bacon in four steps, than if (as it turned out) that number had been three. In other words, the inventors of the Kevin Bacon Game could have picked virtually *any-*

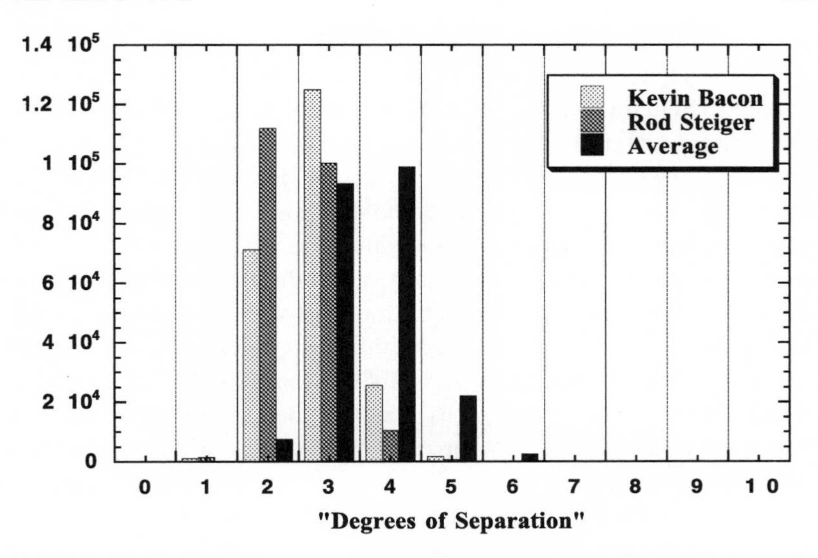

Figure 5.4 Distribution sequences for Kevin Bacon, Rod Steiger, and the average over all actors in the KBG.

one, and they would still have found the same, surprising "centre of the universe" phenomenon. It is perhaps appropriate that in an industry in which lucky breaks make all the difference, Kevin Bacon's new found notoriety could have belonged to any number of other actors, now languishing in his shadow. They too, "could'a been a contender."

5.2 THE POWER OF NETWORKS

It is encouraging that at least one example exists of a real small-world network whose properties can be predicted by the relational-graph model. But nothing in the model (despite its initial motivation) made any assumptions about the type of network to which it could be applied, so the small-world phenomenon probably ought to appear in other sorts of networks also. One class of networks that is (1) completely different from any kind of collaboration network and (2) certainly of inherent scientific interest is that of power grids.

5.2.1 Examining the System

According to the Electric Power Research Institute standard reference for extra-high-voltage (EHV) and ultra-high-voltage (UHV) transmission lines (General Electric Company 1975), the electrical energy output of the United States in 1975 was 1638×10^9 kWh (kilowatt hours) or 8,000 kWh for every person in the country. By 1990 overall electrical energy requirements were forecast to be more than three times their 1975 levels. This massive requirement for electrical energy is satisfied by a nationwide network of power generators, substations, and tens of thousands of miles of high-voltage transmission lines.[12] Consequently the structural integrity, safety, and efficiency of this network is, and will continue to be, of extreme importance. Graph theory has long been an appropriate and useful tool in the study of electrical networks (Lin 1982; Chung 1986; Chowdhury 1989; Erhard et al. 1992), and electrical networks have often been used as real-life examples of graphs (see Chapter 2 of Bollobás 1979, for example). Hence it is natural to consider the power grid of the United States, or a portion thereof, as a single, connected graph and to measure it structural characteristics. As with any large network, the key obstacle in this exercise is obtaining the requisite data in the required graph format. Fortunately this criterion is satisfied by the power-transmission grid for the states west of the Rocky Mountains[13] (see Fig. 5.5). In order to generate from this the *Western States Power Graph* (WSPG), the following assumptions (which, from the perspective of power engineers, may seem egregious) are necessary:

1. All transmission lines are assumed to be bidirectional; hence the resulting graph is undirected.
2. The nodes of the network—in reality generators, transformers, substations, and so on—are treated as identical, featureless vertices.
3. All transmission lines are assumed to be identical (that is, unweighted), ignoring the important fact that the voltage varies considerably (from 345 up to 1,500 kv) and that different lines have significantly different carrying capacities, impedances, and physical construction.
4. Only the transmission network is considered. This ignores an entire (much larger) associated network, responsible for distributing power from the grid to individual houses, offices, factories, and so on.

These assumptions may limit the usefulness of the resulting graph as far as any engineering/dynamical properties may be concerned, and this would be a significant objection to such measures. However, for the purposes of this exercise, dynamical properties are not of primary

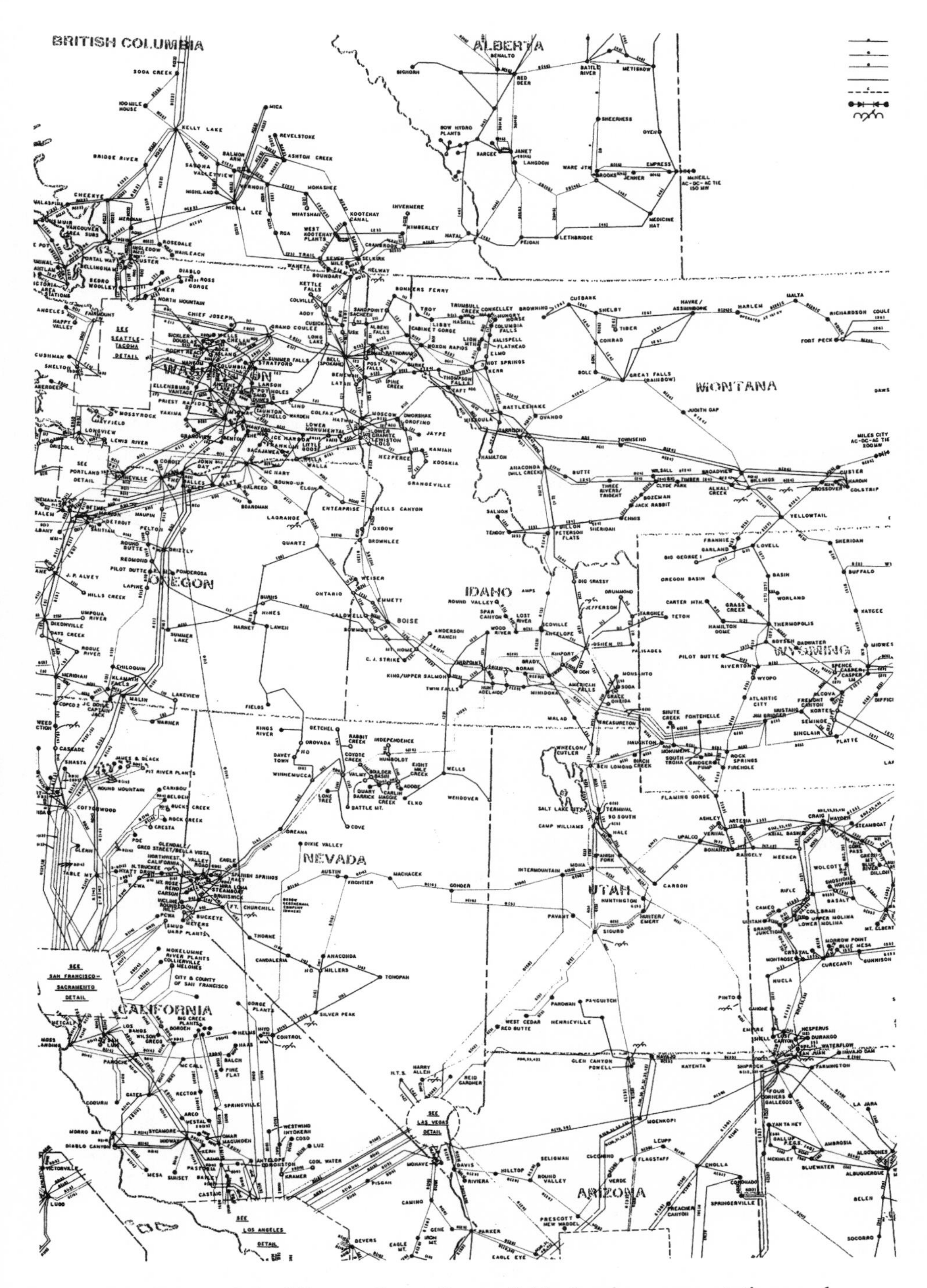

Figure 5.5 A map of the Western States Power Grid, showing power stations and substations as dots (vertices) and high-voltage power transmission cables as lines (edges).

TABLE 5.3
Computed Statistics
for the WSPG.

	Value
n	4,941
k	2.67
L	18.7
γ	0.08
ϕ	0.79
ψ	0.80

concern. What is of interest is a purely topological characterisation of networks that addresses the very general question: Can very different sorts of systems have similar structural features? From this perspective, the above assumptions are quite reasonable because they yield a graph that, in addition to being well defined and completely known, can be analysed using the same framework as was applied to the Kevin Bacon Graph. It is also large ($n = 4,941$) and sparse ($k = 2.67$). In fact, it is sparse almost to a fault, in that it violates the assumption that $k \gg 1$, and so one might expect problems either with connectedness or else in distinguishing clearly between different graph toplogies.

5.2.2 Comparisons

The sparseness of the graph does indeed turn out to be a problem, but some useful observations can still be made. Clearly ϕ and ψ are both much larger than the values considered heretofore as small-world graphs (see Table 5.3). Also, compared to previous results, γ looks "small" and L looks "big."[14] Nevertheless, the graph is still connected, and γ_{WSPG} is still a factor of about 160 times larger than the expected value for an equivalent random graph, whereas L is only about 1.5 times greater. So it appears that the WSPG is a small world after all.[15] The main problem arises with the comparison between the data and the relational-graph model, because the length and clustering approximations break down before ϕ becomes as large as the measured value. This breakdown occurs because the Moore Graph approximation of the local length scale (L_{local}) diverges when $k_{local}(\phi) < 2$. Thus, the curves in Figures 5.6 and 5.7 do not extend all the way to the respective data points. Nevertheless, it is clear, by fitting a curve to the data and extrapolating for large ϕ, that the characteristic path-length prediction of the relational model is

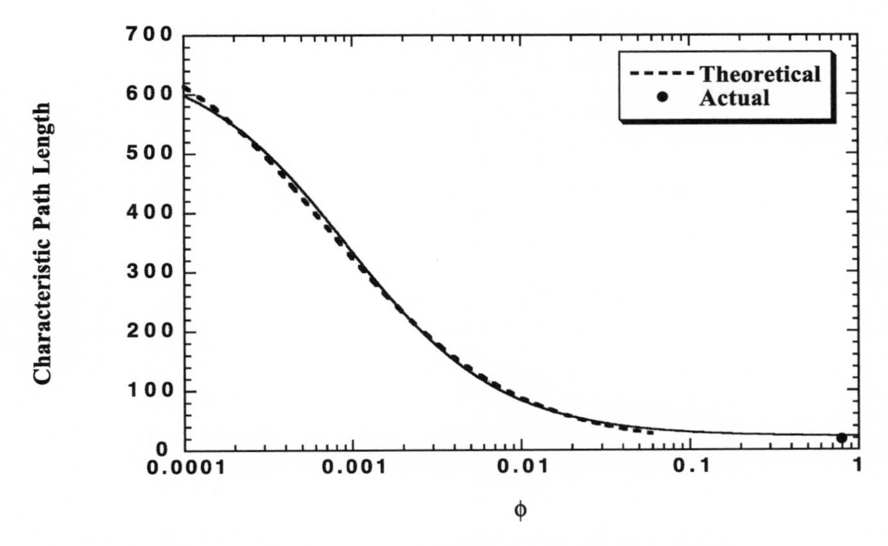

Figure 5.6 Theoretical prediction for $L(\phi)$ vs. ϕ for WSPG parameters ($n = 4{,}941$ and $K = 2.67$) with measured $L(\phi_{\mathrm{WSPG}})$ superposed. As the approximation breaks down for $\phi < (\phi)_{\mathrm{WSPG}}$, a curve is fitted to the predicted data and extrapolated to $\phi \approx 1$.

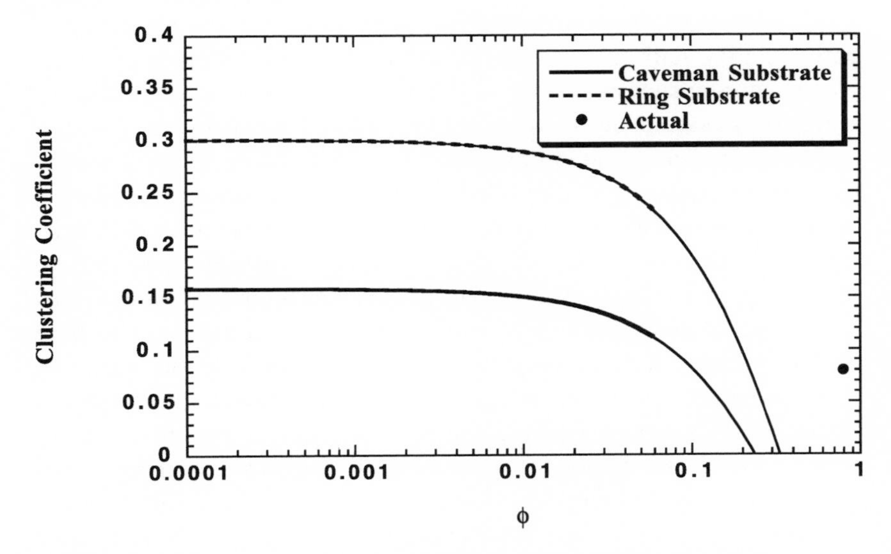

Figure 5.7 Theoretical predictions for $\gamma(\phi)$ vs. ϕ for caveman and ring substrates ($n = 4{,}941$ and $k = 2.67$) extrapolated for large ϕ. Clearly neither prediction matches actual γ_{WSPG}.

TABLE 5.4
Comparison of Computed Statistics for the WSPG with Statistics Predicted by Various Models

	WSPG	Relational	Caveman	1-lattice	Random
L	18.7	22	674	926	12.4
γ	0.08	—	0.65	0.3	0.0005

very much on the right track (Fig. 5.6). Unfortunately, the clustering-coefficient approximation does not do so well, and even the extrapolating curves fail to approach the corresponding data point (Fig. 5.7). Despite these problems, Table 5.4 shows that the relational model still does at least as well as the alternatives. Unlike the Kevin Bacon Graph, it does not much matter whether ϕ or ψ is used as the driving parameter, which actually makes sense, as there is no natural grouping of the power grid vertices in the same way that movies provided natural, fully connected subgraphs of actors in the KBG. Hence one would expect that shortcuts and contractions go more or less hand in hand.

The final comparison to make is with spatial graphs. Once again, there is no way of knowing from the data which ξ (if any) is the natural length scale, so all that can be determined is whether or not *any* ξ can satisfy both L and γ statistics. In this case, because the actual system that the graph represents exists in what is effectively a two-dimensional space, it seems only reasonable to broaden the search criteria further by seeking the appropriate ξ in either a *one- or two-dimensional* spatial model. Despite this added flexibility, Figure 5.8 shows that no such value of ξ exists (although the two-dimensional model is clearly a much closer description), indicating that the relational model provides a superior fit. This really *is* surprising, as one might expect that a physical system, such as a power grid, in the construction of which physical distances have obviously played a role would best be described by a spatial model of the appropriate dimension (in this case, two dimensions). Of course, the spatial model is flawed in that it assumes a uniform, homogeneous distribution of vertices, which is far from the truth. Nevertheless, the *principal* reason for the spatial model's failure appears to be its inability to admit the occasional global edge. Such edges do exist in the WSPG (as can be seen in Figure 5.5 in the form of transmission lines that span whole states), and these are responsible not only for connecting the graph, but also apparently for its small-world properties.

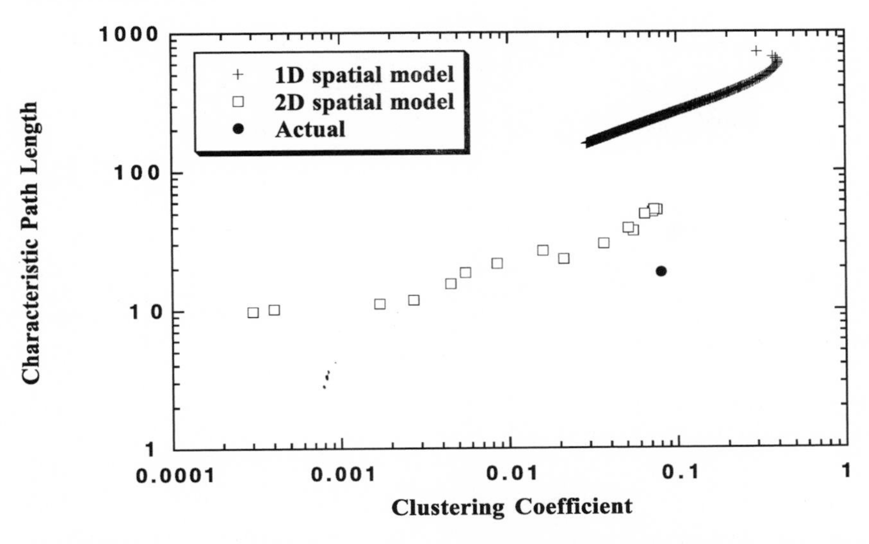

Figure 5.8 Parameterised plot of $L(\xi)$ vs. $\gamma(\xi)$ for the WSPG. One-dimensional data are from analytical, uniform-spatial model, and two-dimensional data are computed numerically. Both data sets describe (γ, L) for all ξ. The actual value computed for the WSPG is also shown.

5.3 A WORM'S EYE VIEW

Occasionally in the history of biological investigation an organism emerges that is set apart by scientists as worthy of special, intensive investigation. These model organisms are living laboratories in which theories can be tested and from which lessons and ideas can be drawn that apply in circumstances far more general and significant than those of the creatures themselves. In genetics it is the famous fruit fly *Drosophila*, in behavioural science it is the chimp, in cancer research the mouse. But there is one organism about which more is known than any other: the tiny worm *Caenorhabditis elegans*, or *C. elegans* for short (see Fig. 5.9), a creature of sufficient notoriety that it even commands its own web site.[16] For over thirty years, ever since Sydney Brenner first convinced the British Medical Research Council that *C. elegans* was a suitable choice of biological benchmark, thousands of scientists have examined, dissected, dissolved, and otherwise interrogated countless numbers of these millimetre-long, transparent, free-living, soil-dwelling nematodes in the search for perfect knowledge of at least one thing in the world. They have come a long way. Every one of its 959 cells has

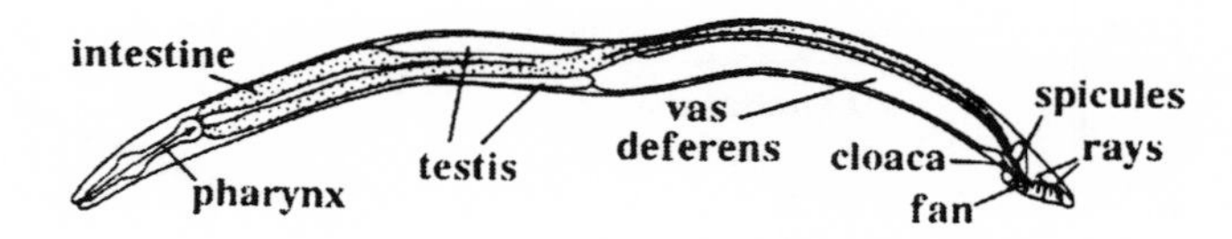

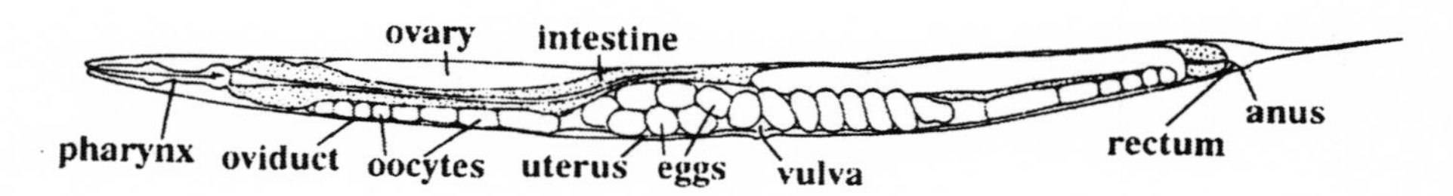

Figure 5.9 Line drawings of *C. elegans.* The male worm is illustrated at top, and the hermaphrodite is shown at bottom.

been mapped out at every stage of its development (Wade 1997). Possibly an even greater triumph is that its entire genome was sequenced in late 1998. Seemingly arcane cellular mechanisms, such as cell death, axon guidance, and cell signalling, were first discovered by worm biologists and have since been found to have significant consequences not only in worms but also in humans. Finally, *C. elegans* possesses a relatively small nervous system (only 302 neurons in its most common, hermaphrodite form), and not only each neuron, but almost every connection between them, has been painstakingly recorded (White et al. 1986). It is this last achievement that will concern us here.

5.3.1 Examining the System

As mentioned above, the neural network of *C. elegans* consists of a paltry-sounding 302 neurons. However, this number is deceptive, as they span a remarkable range of variety and physical characteristics, consisting of 118 distinct classes grouped into nine ganglia, which are in turn located in one of two general units: a nerve ring of twenty cells at the pharynx, and the rest of the neurophil, comprising the ventral cord, a dorsal cord, and four sublateral processes (see Fig. 5.10).

The complete wiring pattern is known for all of these cells except the twenty pharyngeal cells of the nerve ring, but once again the real story is far from simple. Not only are there two classes of connections (synaptic connections and gap junction connections, which join neuron bodies and processes together directly), but the synaptic connections themselves are subclassified as either presynaptic or postsynaptic (corresponding to

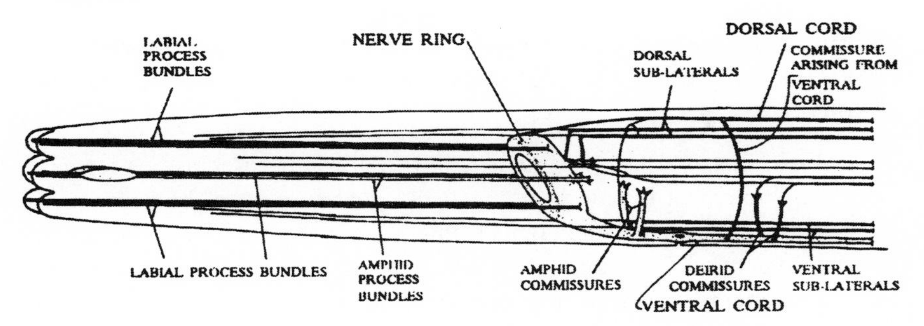

Figure 5.10 The two general units of the *C. elegans* nervous system. A nerve ring of twenty cells at the pharynx is the central region of the neurophil in the worm. The other unit is composed of the rest of the neurophil: a ventral cord (the main process bundle that emanates from the nerve ring), a dorsal cord (axons of motor neurons that originate in the ventral cord and enter the dorsal cord via commisures), and four sublateral processes that run anteriorly and posteriorly from the nerve ring. Reprinted from Achacoso and Yamamoto (1992).

incoming and outgoing *directed* edges). Also there is frequently, but by no means always, a multiplicity of connections between any two neurons (sometimes oppositely directed synapses and sometimes a combination of synapses and gap junctions) and a large variance in the number of connections per neuron. All these factors suggest that the simplistic techniques that have been introduced here are incapable of treating the *C. elegans* neural network in a meaningful fashion. This is probably true, at least if one is seeking a *biologically* meaningful analysis. Specifically, in order to utilise the small-world framework as it currently stands, the following steps would be necessary:

1. Ignore the twenty pharyngeal cells of the nerve ring entirely (as, according to Achacoso and Yamamoto 1992, their connectivity has not yet been sufficiently characterised).
2. Ignore all characteristics of the remaining 282 neurons, treating them as identical.
3. Treat synapses and gap junctions as indistinguishable edges.
4. Ignore edge multiplicity, considering only whether vertices are adjacent or not.
5. Treat all edges as undirected.

Of these, the last is potentially the most offensive to biologists, in that the overwhelming importance of neurons is that they transmit messages. So to ignore the fact that they can transmit them in only one direction is to misunderstand and misrepresent the very point of the network itself. Acknowledging this substantial objection, the following defence can be offered.

1. First, this analysis does not attempt to draw any conclusions regarding biological function, just as it did not attempt to draw any conclusion regarding sociological function for the Kevin Bacon Graph or electrical function for the Western States Power Graph. Rather, its concern is purely the investigation of how networks are connected in the broadest possible sense. Functional applications may follow from this, but each system must then be considered on its own, replete with significant addtional colour and detail. However, if any unifying theme is to emerge at all between such vastly different networks, it must necessarily be of a highly abstract nature.
2. Second, in any case, there may be *some* biological validity to the treatment of neurons as undirected edges after all. Specifically, it is known that axons are responsible for not only the transmission of electrical signals but also the transport of substances essential for metabolism and other cell processes. This axonal transport is thought to arise as a function of the transported cells binding to intracellular tracks via "legs" called *kinesins.* This mechanism allows for (and is observed to support) transport in *both* directions along the interior length of the axon (Edelstein-Keshet 1988, p. 461). Hence, although understanding of the electrical functions of the neural network undoubtedly requires that neurons be treated as unidirectional, the *complete* functioning of the neural system may indeed require us to think of neurons as directed for some purposes and undirected for other.

In any case, justified or not, let us proceed to compute the statistics of the *C. elegans Graph* (CeG) in the same fashion as before, using publicly available data in computer-accessible format (Achacoso and Yamamoto 1992). Table 5.5 lists the results.

5.3.2 Comparisons

As with the previous two examples, the actual, computed statistics of the CeG can be compared with the values they would take if the graph conformed to one of the simple models proposed earlier. The message here is mixed. On the one hand, Table 5.5 demonstrates that the graph is a small-world graph in the sense its characteristic path length is close

TABLE 5.5
Statistics Computed for the CeG

	Value
n	282
k	14
L	2.65
γ	0.28
ϕ	0.07
ψ	0.16

to that of an equivalent random graph, yet its clustering coefficient is greater by an order of magnitude. On the other hand, the relational-graph model does not accurately predict either of the graph's statistics for the measured value of ϕ. Specifically, Figures 5.11 and 5.12 show that both L and γ predicted by the relational-graph model are higher than that measured for the CeG—γ significantly so. In fact, the random-graph model produces results that are neither significantly better nor worse than the relational model, although both do considerably better than the other strawgraphs (see Table 5.6). This result casts doubt on the appropriateness of the relational model for this network, which is perhaps not surprising given how far this example is from the original motivation of the model. In particular, one might question the assumption, in the relational-graph model, that randomly rewired edges can connect vertices drawn from anywhere in the graph, with equal probability. The inherent spatial nature of a neural network would argue against such connections, so one might expect that some kind of spatial model would provide a better description of the graph.

Somewhat surprisingly, this does not seem to be the case, at least not for the simplistic, one-dimensional spatial model proposed in Chapter 2.[17] This yields the unhappy result that *none* of the models proposed—random, ordered, relational, or spatial—appear to capture even the simplest topological features of the *C. elegans* network. In defence of the relational model, however, Figure 5.13 indicates that it can be made to satisfy the L and γ statistics of the real graph, if *any* value of ϕ is allowed. That is, the relational-graph model could explain the observed length and clustering characteristics of *C. elegans* if only ϕ were much higher than actually measured. It is possible, even likely, that the principal reason for the model's failure lies in its assumption of large n—a criterion that the CeG graph does not really satisfy (recall that

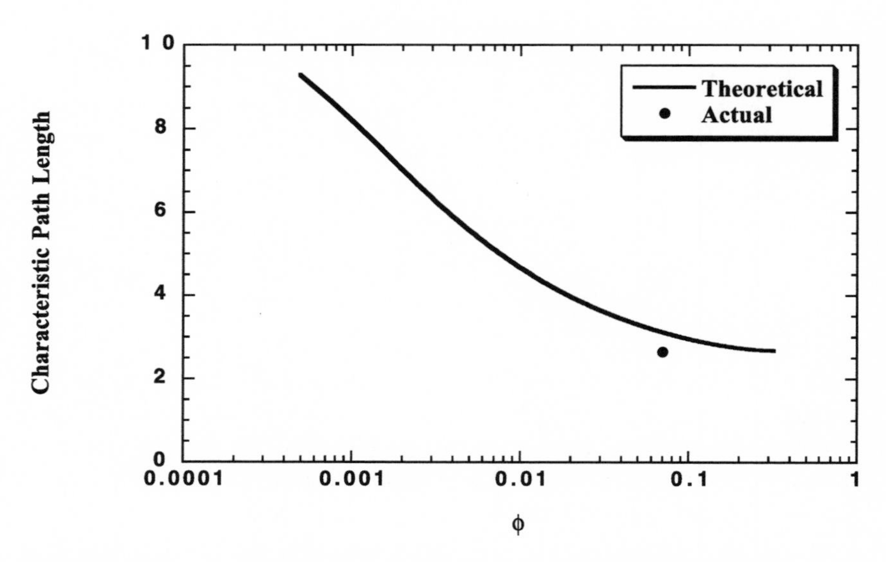

Figure 5.11 Theoretical prediction for $L(\phi)$ vs. ϕ for CeG parameters ($n = 282$ and $k = 14$) with actual $L(\phi_{\mathrm{CeG}})$ superposed.

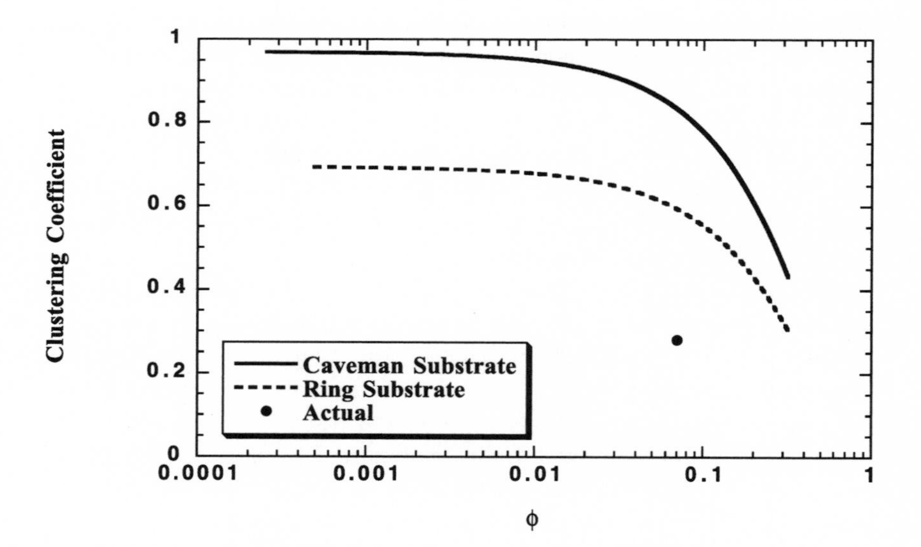

Figure 5.12 Theoretical predictions for $\gamma(\phi)$ vs. ϕ for CeG parameters ($n = 282$ and $k = 14$) with actual $\gamma(\phi_{\mathrm{CeG}})$ superposed. As with the KBG, predictions for both ring and caveman substrates are shown.

TABLE 5.6
Comparison of Computed Statistics for the CeG with Statistics Predicted by Various Models

	CeG	Relational	Caveman	1-lattice	Random
L	2.65	3.1	11	10.5	2.25
γ	0.28	0.59–0.83	0.98	0.69	0.050

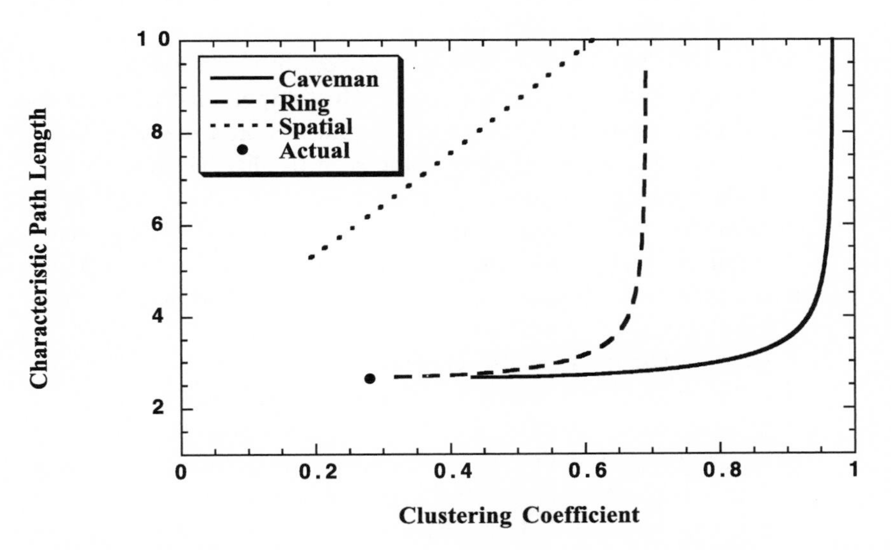

Figure 5.13 Paramterised plot of $L(\xi)$ vs. $\gamma(\xi)$ for relational and uniform-spatial graphs with *C. elegans* parameters. Actual CeG value is also shown.

the smallest graph considered up to this point was $n = 1{,}000$, $k = 10$). What is still encouraging is that the data point $(\gamma_{\mathrm{CeG}}, L_{\mathrm{CeG}})$ appears to lie *somewhere* on the curve defined by the relational-graph model, indicating that a model of this *sort* (one with occasional, very long range edges) is at least on the right track and may have better success with larger neural networks.

5.4 OTHER SYSTEMS

The three systems explored above were chosen in part because of their inherent interest, in part because they satisfied (to varying extents) the required conditions for consideration as small-world graphs, but largely because they were available in a format that could easily be put into a computer as the adjacency matrix of a graph. What follows is something of a wish list of graphs whose definitions can be specified and that might also exhibit interesting properties, but that have not yet been mapped out. The purpose of listing them is that perhaps someone else will be prompted to do the requisite work and tell us what they find.

1. *The Collaboration Graph*. Defined at the start of the chapter—obviously of interest to a great many people and widely believed to be a small-world graph.
2. *Science Citations*. Define a vertex as a scientific paper and a *directed* edge as a citation of another paper. Most papers' list of citations are from a relatively small subset of journals, and many of those papers will cite, or be cited by, others on the list. However, occasionally a paper will appear that draws together work from many different fields, proposes an idea that is subsequently used in many different fields, or both. Such a graph, which would represent the development of scientific ideas, may or may not harbour the characteristics of a small world, but it seems plausible that it might.
3. *Word Associations*. Define each word as a vertex and connect vertices by edges if two words are "associated with" each other (for instance, "bow and arrow" but also "bow and ribbon" or "bow and hair" or "bow and ship" or "bow and scrape"). Other definitions of edges might be "sounds like" or "looks like" or even "shares a meaning with." If such graphs could be constructed, they might well provide some stimulating insight into the workings of language and the way that a sequence of apparently disparate ideas can appear, in rapid succession, in the flow of a conversation or a train of thought.
4. *Organisational Networks*. Individuals are vertices, and edges represent working or personal relationships. Such a construction might be useful for modelling the flow of information around an organisation or the emergence and sustenance of cooperation (see Chapter 8 for a beginning).
5. *World-Wide Web Links*. Define each web page as a vertex and each of its hotlinks (to other web pages) as edges. Despite the millions of web pages and obvious sparseness of the corresponding graph, the author bets that almost any web page can be reached from almost any other within $O(10)$ links.

Furthermore, there is nothing about the definitions of characteristic path length, clustering, shortcuts, or any other statistics that precludes their application to directed or weighted graphs. Hence even the graphs already considered in this chapter should probably be revisited and analysed in greater detail, once the commensurate embellishments of the relevant theoretical models have been completed.

5.5 MAIN POINTS IN REVIEW

It is difficult to say exactly how the results of this chapter should be interpreted, but an optimistic appraisal leads to some remarkable and exciting speculations. Three real (albeit somewhat sanitised) graphs are examined, spanning three orders of magnitude in n and ranging across the sciences from sociology to biology to engineering. In each case, the actual characteristic path length and clustering coefficient is measured and compared with the predictions of all the contending explanatory models. In summary, two statements can be made, in answer to the two questions at the start of the chapter.

1. Each graph exhibits a characteristic length comparable to the smallest length achievable for a graph of that size (as attained by a random graph), but the clustering coefficient is much greater than we would expect for an equivalent random graph. Hence all the graphs considered are *small-world graphs*. This result implies that small-world properties are not just an interesting consequence of theoretical constructions but actually exist in real networks that are partly ordered and partly random.
2. For the Kevin Bacon Graph and the Western States Power Graph, the relational-graph model provides the best fit to the observed data of any of the proposed models. For the *C. elegans* Graph, all models perform poorly. This is not so surprising, given how removed the *C. elegans* Graph is from the abstract foundations of the relational-graph model, and given also that n is probably not large enough for the model to generate reliable statistics. However, conceding that ϕ is a crude statistic (especially for small n), then there is evidence that this *sort* of model (as opposed to a spatial model, for instance) might still be useful in understanding the structure of larger neural networks.

Given all the assumptions that were made about the systems in question, it would be unrealistic to relate these findings to any *functional* properties of the actual systems. Furthermore, two of the three graphs in question do not strictly comply with the requirements $n \gg k \gg 1$ upon which the models are based, so the statistics are not reliable enough to justify any truly bold claims. Nevertheless, *something interesting does appear to be going on*. There *does* appear to be a common thread linking these systems together in terms of the qualitative arrangements of their connections, and regardless of its consequences, that is a remarkable thing in itself.

With the question So what? ringing in our ears, it is now time to examine the dynamical consequences of what has so far been an entirely structural phenomenon.

Part II Dynamics

6

The Spread of Infectious Disease in Structured Populations

The sole concern of Part I is the understanding of network structure: how to describe all kinds of networks using models of graphs, which of these exhibit particular, interesting properties, and what mechanisms are responsible for bringing these properties about. Specifically, the focus has been on models of graphs that interpolate, by virtue of a random-rewiring mechanism, between ordered, lattice-like graphs and random graphs. In contrast, the next few chapters deal with the second question posed in the introduction: *Does any of this really matter?* That is, if a large system of dynamical entities is coupled together in a fashion described by some graph, then can random rewiring of that graph have any bearing on the *dynamical* properties of the system? The answer to this is not obvious. Recall that throughout Part I, all comparisons were made between graphs with the same average degree (k). Thus random rewiring cannot lead to comparisons between, say, one-dimensional nearest-neighbour $(k = 2)$ graphs and complete $(k = n - 1)$ graphs. In that case differences in the behaviour of the corresponding dynamical systems would not be surprising, because elements in the complete graph have much more "information" available to them than do their counterparts in the one-dimensional graph. Recall also that many of the most dramatic changes in the structural statistics of the graphs occurred as a result of the rewiring of only a small fraction of edges. So the distinctions are considerably more subtle even than between lattice-like graphs and random graphs. In fact, it turns out that the difference between a big and a small world may not even be detectable at the local level. Can the same thing be true of the dynamics? In other words, can small rearrangements in the coupling network of a distributed dynamical system cause large changes in its corresponding global dynamical properties? This is the question that dominates Part II.

The general procedure is very much like that used in Part I. A number of simple models of dynamical systems will be introduced, each one of which "lives" on a graph. In other words, the vertices of the graph are replaced with elements that exhibit their own, internal dynamics and the edges with coupling relationships between the elements. The graph can then be tuned according to one or more of the models from Part I, and the resulting dynamics measured as function of the known structural parameters. This sort of experimentation should shed light on two related questions:

1. How do the dynamical properties of the system depend upon the structural properties of the *coupling graph*? Specifically,
   a. Does the attractor of the system change? and
   b. If not, does the characteristic transient time to the attractor change?
2. If either of these dependencies exists, is the functional relationship between structure and dynamics understandable?

This last question is clearly an ambitious one. Certainly it is too ambitious for this book despite the fact that only simple abstract models will be considered. Perhaps the simplest such model, and certainly one that bears an obvious relevance to networks, is that of disease spreading.

6.1 A BRIEF REVIEW OF DISEASE SPREADING

In general, what is meant by a *spreading problem* is a spatio-temporal description of an *influence* that propagates through a medium or population after being initiated (by some action outside the system) at a specific point or set of points in the system. All sorts of influences may spread, from diseases in populations through fires in forests, heat in metal bars, and even levels of electrical excitation in heart muscle. This chapter is concerned solely with *disease spreading*, although this can often serve as a metaphor for other influences (like fashion, rumours, or even crime; see Gladwell 1996).

The classical mathematical approach to disease spreading either ignores population structure altogether or treats populations as spatially distributed in a continuous medium. In the first case, the entire population consists of subpopulations (typically "susceptible," "infectious," and "removed") whose number, size, and level of interaction determine the transmission of disease. In graph language, the population is the graph, the vertices are differentiated by subpopulation, and the connectivity is random in that the subpopulations interact in proportion to their sizes. This approach has been utilised effectively in the modelling of infection in well-mixed populations (May and Nowak 1994; Murray 1993, Chapter 19) and also in the human immune system (Nowak and Bangham 1996) where, in both cases, the emphasis is on the detailed dynamics of disease transmission rather than the relationships among subpopulations. The second classical approach introduces a spatial dependency to the subpopulations involved and is typified by reaction-diffusion equations, in which travelling waves and spiral waves are often sought as solutions (Murray 1993, Chapter 20). Here questions of the stability of equillibria and the analytic tractability of solutions tend to dominate.

More recently a third approach has been developed that takes greater account of the fact that populations are often inherently discrete and exhibit high levels of structure, both spatial and social. Kareiva (1990) reviews a number of such attempts (which he calls "stepping-stone" models) in the context of population dynamics and May and colleagues (Hassell et al. 1994; May 1995) have considered various parasite-host problems on two-dimensional grids of discrete but homogeneous patches. Both authors conclude that the introduction of spatial structure to the coexistence of subpopulations and the rate of migration between them can significantly affect both the meta-population size and its susceptibility to parasites and disease. This approach has been used by Hess (1996a, 1996b) to compare virus transmission among subpopulations that are connected according to various simple topologies such as a ring and a star. Also, Sattenspiel and Simon (1988) have considered a detailed model of the spread of an infectious disease in a meta-population in which the coupling topology of the subpopulations is varied. Finally, Longini (1988) has used actual airline routing data as the graph of connections between major cities around the world in order to model the spread of an infectious disease. For the appropriate parameters, he compares his results to the available data concerning the 1968 outbreak of influenza, which initiated in Hong Kong and spread to fifty-two major cities.

None of these various approaches, however, treats the problem of a disease spreading in a population as a *function* of the population structure. The closest is the stepping-stone approach, which deals with a small number of populations, within each of which mixing is always assumed to be random. Also, the models of Sattenspiel, Simon, and Hess do consider different types of connectivity between the subpopulations, but they consider only the extreme cases such as randomly connected or connected in a ring. Furthermore, they do not necessarily maintain a constant vertex degree across the topologies under comparison. A reasonable defence for considering only a few, quite distinct coupling topologies is that the extremal cases are natural to consider because the connectivity of social systems is generally unknown. However, this is, in a way, precisely the point—*because* the connectivity is unknown, it is all the more important to consider as broad a range of coupling topologies as possible to see which characteristics emerge as the dominant influences on disease dynamics.

Some work along these lines has been done recently by Kretzschmar and Morris (1996), who showed that increasing the fraction of *concurrent* partnerships in a population can greatly accelerate the spread of a sexually transmitted disease throughout a network, in which k is fixed at $k = 1$. They *do* consider the dynamics on a one-parameter family of networks, but it is a quite different interpolation to that considered here. In their case, the structural mechanism responsible for accelerating the spread of the disease is the appearance of larger connected components as the constraints on concurrency are relaxed. In contrast, all the graphs considered here are connected and $k \gg 1$, so any changes observed must result from more subtle changes in structure than connectedness.

6.2 ANALYSIS AND RESULTS

6.2.1 Introduction of the Problem

In order to emphasise the importance of network structure, a deliberately simplified model is used. The relevant terminology follows that of Murray (1993, p. 611), who defines three categories into which any member of the population might fall: *susceptible*, *infectious*, and *removed*. Susceptible members are effectively virgin territory for the disease, whilst infectious members are both infected and capable of infecting others with whom they are in *direct* contact. The disease having run its course, infectious members are removed from the population, either temporarily (after which they become susceptible) or permanently (representing either immunity or death). The partial fractions of the population in each category vary as a function of time and are denoted S, I, and R, respectively, where $S(t) + I(t) + R(t) = 1$ always. Additionally, the characteristic times τ_I and τ_R specify the duration (in dimensionless units of time) of the infectious and removed stages of the disease, respectively. Finally, the *infection rate* $0 \leq \rho \leq 1$ is the probability that any one infectious element will infect any one of its susceptible neighbours.

This simple formulation yields essentially two types of dynamics in which the disease (presumably introduced by some outside entity) starts at a single, randomly selected member of the population. The first, in which $\tau_R = \infty$, is called *permanent-removal dynamics* and is analogous to the spread of a lethal and contagious disease. The second, in which $\tau_R < t_{\max}$ (the length of time for which the simulation is run), is called (imaginatively enough) *temporary-removal dynamics* and is reminiscent of the dynamics of excitable media or maybe the dynamics of the influenza virus over a timescale of many years.

6.2.2 Permanent-Removal Dynamics

The general problem, as described above, depends upon three independent parameters: τ_I, τ_R, and ρ. However, for permanent-removal dynamics, one parameter (τ_R) is immediately removed, leaving two that turn out to be dependent. This can be seen by performing a local analysis of the dynamics. Straightforward enumeration yields the expected number of vertices that will be infected directly by an infected vertex v over the course of its infectiousness:

$$E(\text{infected}) = k\left[1 - (1-\rho)^{\tau_I}\right],$$

where $k \gg 1$ is the average degree of the graph. Hence there exists a change of variables $\rho \rightarrow \rho'$, $\tau_I \rightarrow \tau_I'$ such that

$$k\left[1 - (1-\rho')^{\tau_I'}\right] = k\left[1 - (1-\rho)^{\tau_I}\right]. \tag{6.1}$$

Setting $\tau_I' = 1$, then Equation 6.1 becomes

$$\rho' = 1 - (1-\rho)^{\tau_I}. \tag{6.2}$$

Hence, given any ρ and τ_I, ρ' can be chosen according to Equation 6.2 such that the corresponding $\tau_I' = 1$. Because this result is based on a strictly local analysis, it is valid for arbitrary topology of the population. Figure 6.1 confirms this, showing a comparison of $S(t_{\max})$ between $\tau_I' = 1$ and $\tau_I = 10$ on both a 1-lattice and a random graph, where $\rho(\tau_I = 10)$ has been rescaled according to Equation 6.2. So, from this point on, it is sufficient to consider only the special case of $\tau_I = 1$, in which case ρ becomes the only parameter in the problem.

The natural case to consider first is that of a disease spreading from a single source in a randomly mixing population. In terms of the β-model

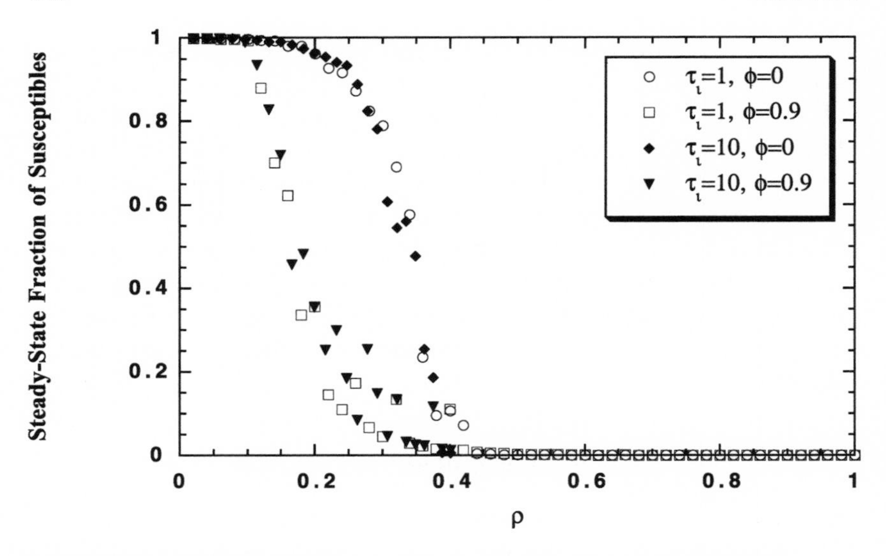

Figure 6.1 Comparison of $S(t_{\max})$ vs. ρ (with $\tau_R = \infty$) for two values of τ_I ($\tau_I' = 1$, $\tau_I = 10$), showing that a rescaling in ρ renders the results equivalent.

of Part I, this is equivalent to $\beta = 1$, or $\xi = O(n)$ for the uniform spatial model. For $\tau_I = 1$, $\tau_R = \infty$, and assuming that population elements susceptible to infection at each time step are independent of those susceptible at any other time step (this is the random-mixing condition), then the total expected number of infected elements in the population at time $t > t_0$ is

$$\begin{aligned} I(t) &= I(t_0) + I_{\text{new}}(t_0+1) + I_{\text{new}}(t_0+2) + \cdots + I_{\text{new}}(t) \\ &= \rho \cdot k + [\rho \cdot k \cdot \rho(k-1)] + [\rho \cdot k \cdot \rho(k-1) \cdot \rho(k-1)] \\ &\quad + \cdots + \rho^t \cdot k(k-1)^{t-1} \\ &= \frac{k}{k-1}\left[\frac{(\rho(k-1))^{t-1}-1}{\rho(k-1)-1} - 1\right]. \end{aligned} \tag{6.3}$$

This expression, however, is valid only for $\rho(k-1) \neq 1$. For $\rho(k-1) < 1$, $I(t) \to 0$ for increasing t and for $\rho(k-1) > 1$, it increases exponentially. At $\rho(k-1) = 1$, however, $I(t)$ increases, but only linearly. Hence the *tipping point* of the disease occurs at $\rho(k-1) = 1$, the value of ρ beyond which the disease explodes exponentially into the susceptible population. This condition yields a critical value of the infectiousness, which defines the tipping point:

$$\rho_{\text{tip}} = \frac{1}{k-1}. \tag{6.4}$$

In practice, the tipping point is defined somewhat differently: the value of ρ at which an initially infected population of asymptotically negligable size ($o(n)$ as $n \to \infty$) grows to infect a population on the order of the entire population. Nevertheless, in a randomly mixing system, these two definitions should coincide,[1] and Figure 6.1 shows that for $k = 10$, $\rho_{tip} \approx 0.11 \approx 1/9$.

Having gained some understanding of the simplest case, the next step is to compare some results for an entire range of topologies as a means of answering the two questions posed at the start of this chapter, at least within the context of this simple model. Figure 6.2 shows the steady-state fraction of susceptibles $S(t_{max})$ versus ρ for three values of ϕ, where β-graphs ($n = 1{,}000$, $k = 10$) have been used to determine the couplings of the system. Three distinct regions are apparent in Figure 6.2:

1. For $\rho < \rho_{tip} \approx 0.11$, all topologies (that is, all values of ϕ) yield the same result: the disease infects only an $o(n)$ population before dying out. This is a *trivial steady state* in the sense that nothing happens that can distinguish between different topologies.
2. For $0.11 \lesssim \rho \lesssim 0.5$ different topologies yield different $S(t_{max})$.
3. For $\rho \gtrsim 0.5$, all topologies once again yield the same end result, only this time it is a *nontrivial steady state* because the disease has taken over the entire population.

There is nothing more to say about region 1, but regions 2 and 3 deserve some extra attention. Region 2 is confusing, and it is not clear what the functional relationship is between structure and dynamics. That there is *some* significant relationship is clear. However, cross sections (for fixed ρ) through Figure 6.2 reveal that the dependence of $S(t_{max})$ on ϕ changes dramatically for different ρ. For some values of ρ, the dependence is reminiscent of the clustering coefficient from Part I (Fig. 6.3); for other values of ρ, it is equally reminiscent of the characteristic path length (Fig. 6.4); and for others still, it is reminiscent of neither.

Figure 6.5 demonstrates that the results just stated do not depend upon *which* model of relational graph (α or β) is used as long as ϕ is used as the driving parameter. Hence the dynamics of this model are largely independent of the difference between the alternative relational-graph models and dominated by the same factors that dominate the structural statistics of Part I.

All these results can be summarised in a single figure by plotting the tipping point (ρ_{tip}) as a function of ϕ (Fig. 6.6). This is somewhat prob-

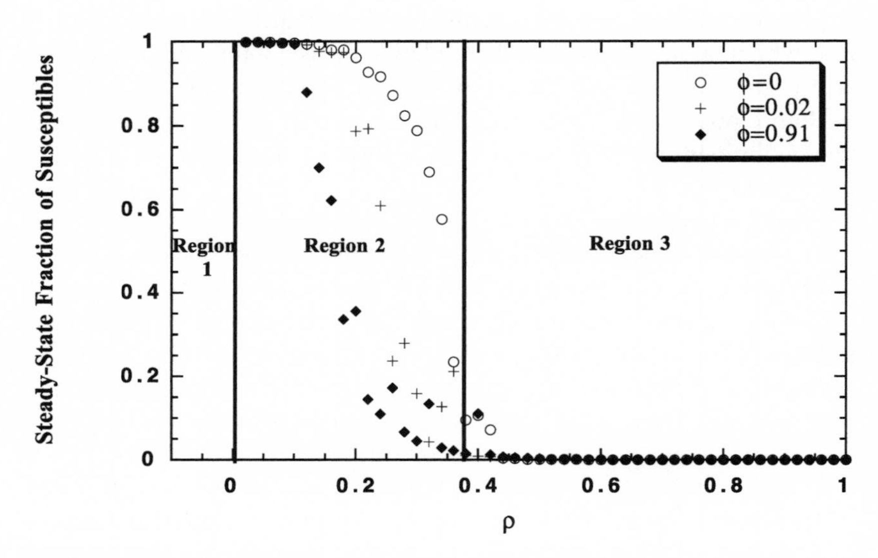

Figure 6.2 $S(t_{\max})$ vs. ρ for permanent-removal dynamics ($\tau_R = \infty$), for three values of ϕ. ($n = 1{,}000$, $k = 10$). Changing has different effects in the three regions.

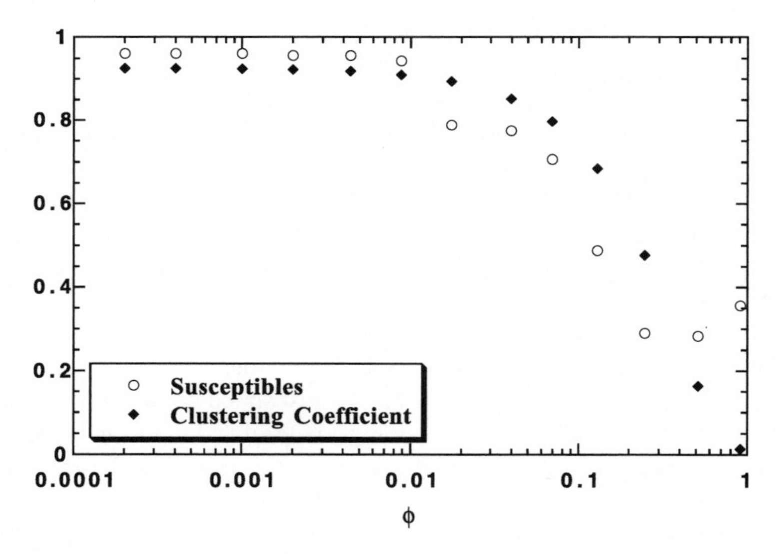

Figure 6.3 $S(t_{\max})$ vs. ϕ (for $\rho = 0.2$) compared with a scaled version of $\gamma(\phi)$ for an equivalent β-graph ($n = 1{,}000$, $k = 10$).

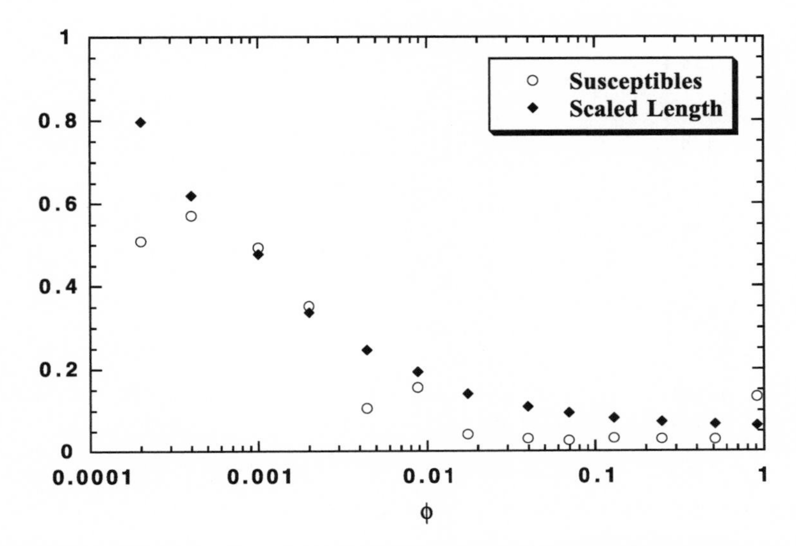

Figure 6.4 $S(t_{\text{max}})$ vs. ϕ (for $\rho = 0.32$) compared with the scaled version of $L(\phi)$ for an equivalent β-graph ($n = 1{,}000$, $k = 10$).

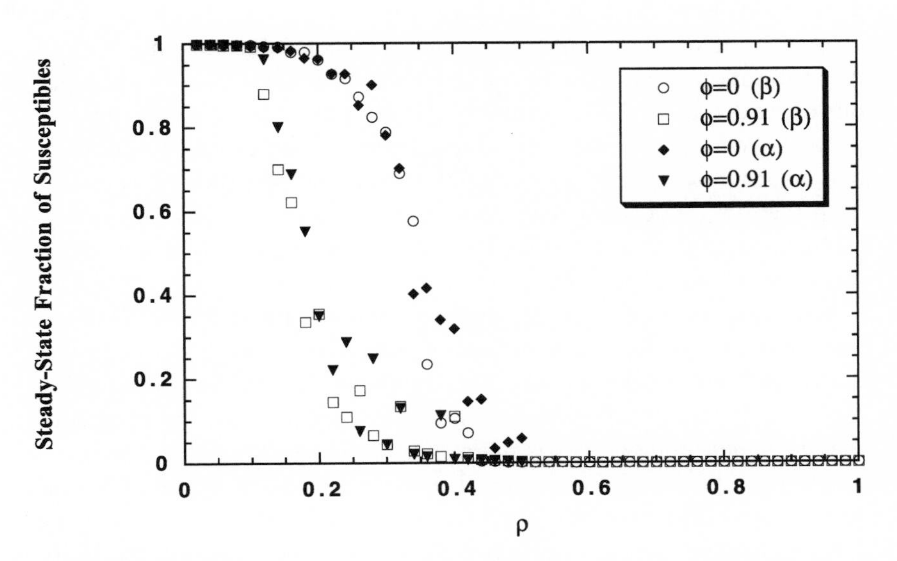

Figure 6.5 Comparison of $S(t_{\text{max}})$ vs. ρ (with $\tau_R = \infty$) for α- and β-graphs.

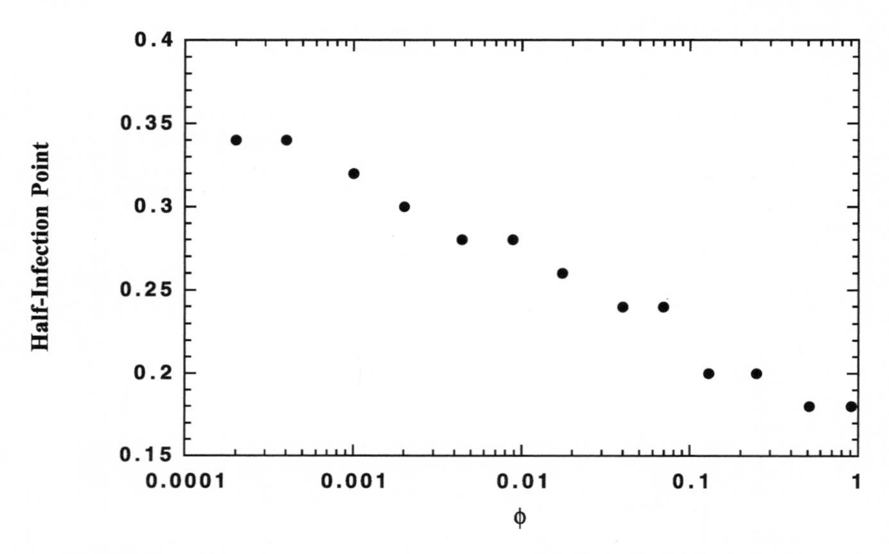

Figure 6.6 Half-infection point (ρ_{half}) vs. ϕ for permanent-removal dynamics with β-graph coupling.

lematic in practice, as for small ϕ, $S(t_{\max}, \rho; \phi)$ meanders from high to low values rather than dropping precipitously as the random-mixing case does. To avoid all this, the *half-infection point* ρ_{half} can be defined as the value of ρ at which *half the population* has been infected. This is more of a saturation value than a tipping point, but it exhibits a less ambiguous dependence on ϕ. Nevertheless, this does not really clear up matters as far as relating structure to dynamics goes, because $\rho_{\text{half}}(\phi)$ does not bear much resemblance to *any* of the structural statistics of Part I or even the $S(t_{\max})$ dependencies above. All are monotonically decreasing, and ρ_{half} does decrease rapidly for small ϕ, but beyond that, not much can be said.

Nevertheless, the broader of the two questions posed above has been answered, at least for this dynamical system: the attractor for the global dynamics really does depend, often quite sensitively, on the coupling topology. In epidemiological terms, this is equivalent to the statement that *the tipping point of the disease can be highly sensitive to the connective topology of the population*. In particular, Figure 6.6 indicates that epidemics can occur at lower ρ than is the case in, say, a 1-lattice. This has implications for the way we think about social diseases, which are often perceived as confined to isolated subgroups of a population. The message here is that the highly clustered nature of small-world graphs

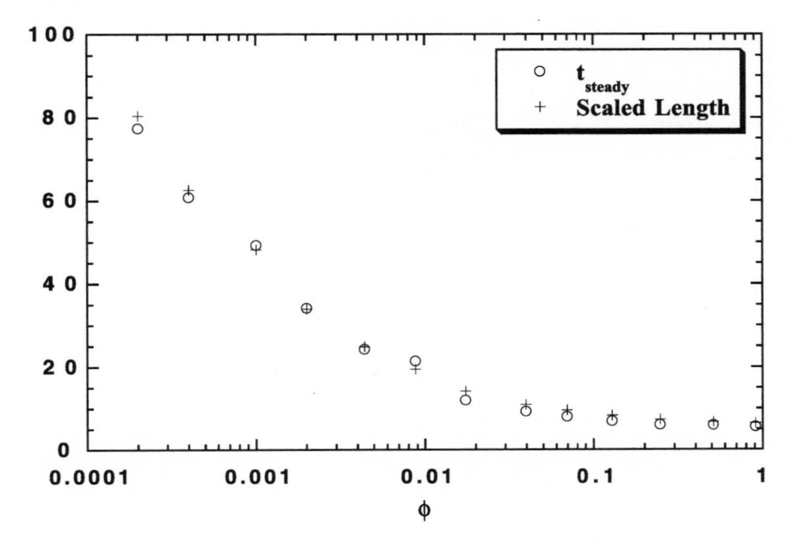

Figure 6.7 $t_{\text{steady}}(\rho = 1)$ vs. ϕ compared with twice $L(\phi)$ for an equivalent β-graph ($n = 1{,}000$, $k = 10$).

can lead to the intuition that a given disease is "far away," when, on the contrary, it is very close. The fact that so few shortcuts may be required to achieve this small-world effect is important, because such tiny alterations to the network configuration could be impossible to detect from the perspective of an individual.

Region 3 is a somewhat simpler matter to examine, as the disease always takes over the entire population regardless of its connective topology; it is merely a question of how long this takes. However, Figure 6.7 shows that in this region the *time taken* to reach a steady state (t_{steady}) varies dramatically as a function of ϕ and that disease can spread on a small-world graph far more rapidly than in a 1-lattice and almost as fast as in a random graph. Not surprisingly, given the close analogy between spreading dynamics and the distribution sequence defined in Chapter 2, $t_{\text{steady}}(\phi)$ bears a close functional relationship to $L(\phi)$. In fact, the case of $\rho = 1$ can be understood in graph-theoretic terms as the *eccentricity* of the vertex from which the disease spreads, where the eccentricity of a vertex (i) is the maximum $d(i, j)$ over all other vertices (j) in the graph. Because, for any i, $\max(d(i, j))$ must be at least as large as the *average* $d(i, j)$, and because the *diameter* D of a graph is just the maximum eccentricity over all vertices (i), then t_{steady} is necessarily bounded as follows:

$$L \leq t_{\text{steady}} \leq D.$$

For most graphs, the two statistics L and D are related by a constant factor, although this will be different for different topologies. For instance, in the case of a 1-lattice, $D = 2L$, whereas for a random graph, in which the distribution sequence grows exponentially, it follows that most vertices will be separated by distances close to the diameter, and so $D < 2L$. Hence across the full spectrum of β-graphs (or uniform spatial graphs) it must be true that

$$L \leq t_{\text{steady}} \leq 2L. \tag{6.5}$$

Figure 6.7 supports this result, although only approximately as t_{steady} is averaged over only a subset of the entire graph. In this case, then, the transient time for the dynamics to reach the same, nontrivial steady state bears a relatively simple and obvious relationship to the structure of the underlying graph; that is, shorter characteristic path length implies faster spreading of the disease. In a real-world scenario, where an epidemic can be responded to, the timescale on which it spreads becomes a crucial factor.

Corresponding results can also be generated for spatial graphs, except where ξ is used instead of ϕ for the driving parameter. Little qualitatively new happens here in that the parameter space can be divided into the same three regions as before (see Figure 6.8). The only significant difference is that there appears to be somewhat less variation in the functional dependence of $S(t_{\text{max}})$ on ξ in region 2 than for relational graphs. Figure 6.9 shows that, for uniform spatial graphs, the most dramatic changes in $S(t_{\text{max}}, \xi; \rho)$ occur for small ξ, during which the asymptotic value is approached. Figure 6.10 shows that, unlike relational graphs, ρ_{half} for the spatial model is functionally similar to $L(\xi)$. Perhaps the reason for this is that $S(t_{\text{max}})$ and ρ_{tip} depend in general on both γ and L. As noted in Chapters 3 and 4, γ and L have strikingly different functional forms for relational graphs, but very similar functional forms for spatial graphs. It is perhaps not surprising, then, that the dependence of dynamics upon structure is more complicated for relational graphs.

6.2.3 Temporary-Removal Dynamics

Once $\tau_R < t_{\text{max}}$, all three parameters in the model (along with the graph parameter ϕ or ξ) must be retained, and the dynamics become corre-

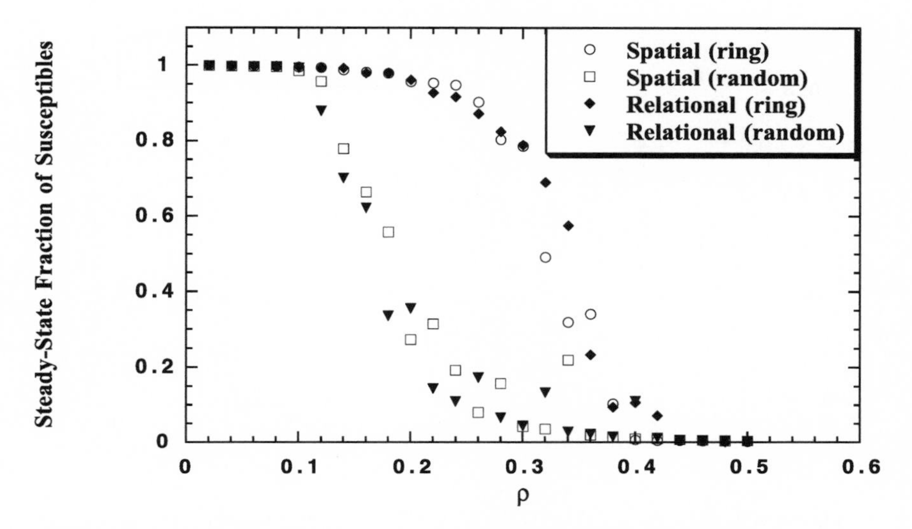

Figure 6.8 Comparison of $S(t_{\max})$ for permanent-removal dynamics, between relational and spatial graphs at their respective extremes.

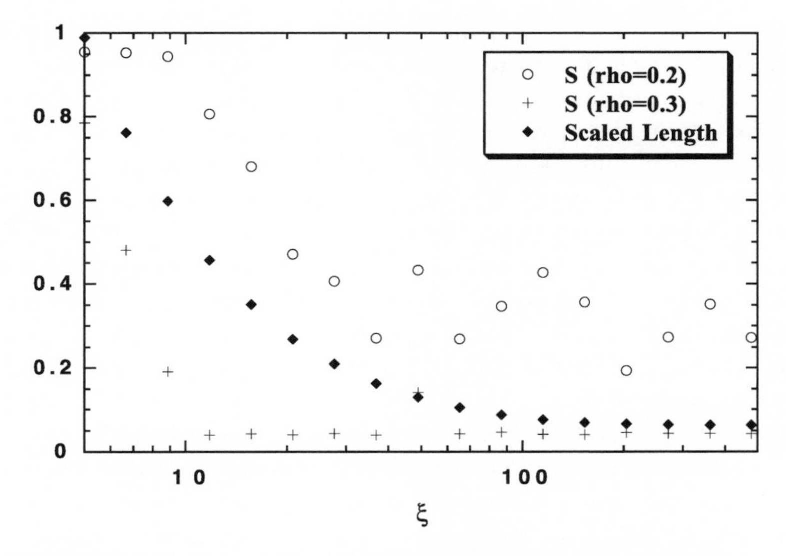

Figure 6.9 $S(t_{\max})$ vs. $\xi(\rho = 0.2, 0.3)$ compared with a scaled version of $L(\xi)$ for an equivalent uniform spatial graph ($n = 1{,}000$, $k = 10$).

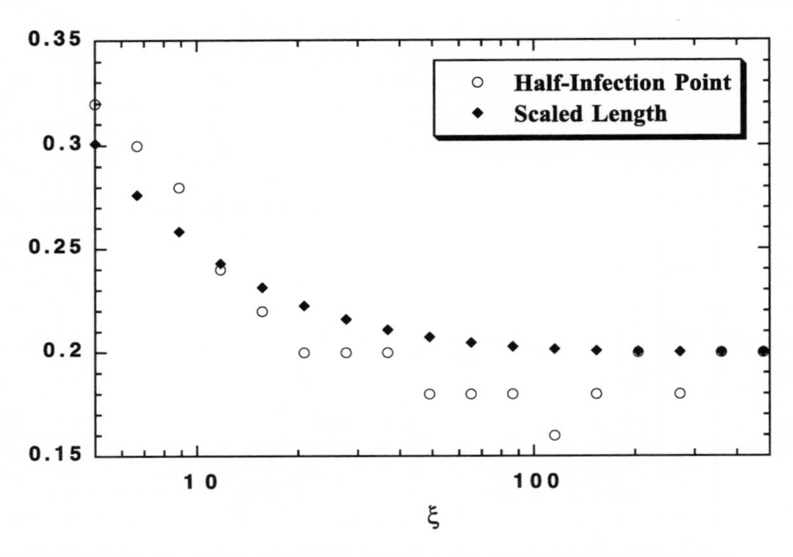

Figure 6.10 ρ_{half} vs. ξ compared with a scaled version of $L(\xi)$ for an equivalent uniform spatial graph ($n = 1{,}000$, $k = 10$).

spondingly harder to analyse so not much will be said about this case. Here just one set of parameters ($\tau_I = \tau_R = 1$, and variable ρ) is examined, in order to show that the flavour of the results obtained above is preserved even for more complicated situations. The dynamics of the temporary-removal case does not tend to either of the permanent-removal steady states, where the disease either dies out or takes over the entire population. Rather, individual elements continually fluctuate between the three stages of disease. Nevertheless, the partial fractions $S(t)$, $I(t)$, and $R(t)$ *do* approach asymptotic values, whose fluctuations are small compared to the $O(1)$ fluctuations they make during their transients. Hence it is still possible to examine the asymptotic partial fractions as a function of the graph topology and, in cases where the same attractor is approached, examine the effect of topology upon the characteristic transient time.

Figure 6.11 shows that, for relational graphs, the asymptotic value of $S(t_{\text{asymp}})$ exhibits a very weak dependence upon ϕ. This is an interesting contrast to the permanent-removal dynamics, in which ϕ was a significant determinant of the steady-state dynamics and further complicates any attempt to draw general relationships between structure and dynamics. Nevertheless, the dynamics is not independent of structure. As Figure 6.12 demonstrates, the characteristic transient time varies in strict

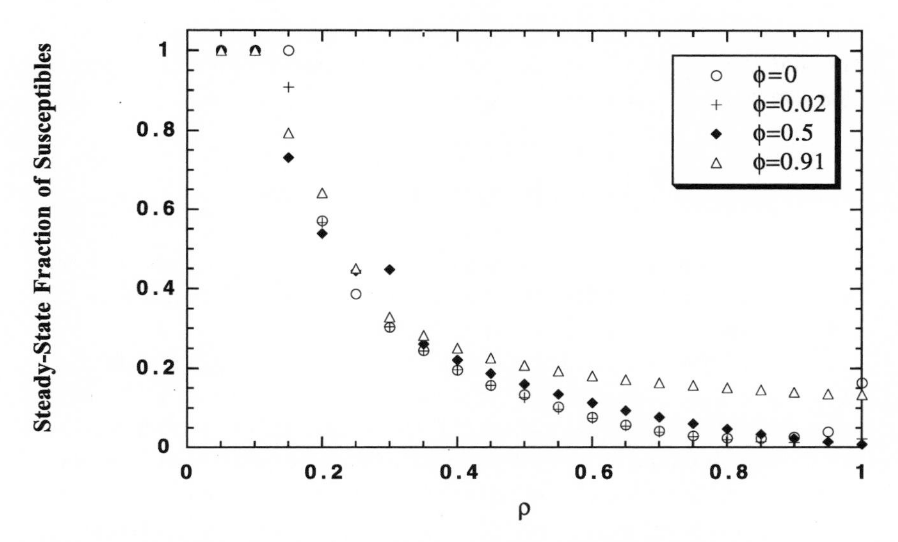

Figure 6.11 Asymptotic value of S vs. ρ for temporary-removal dynamics, on β-graphs with four ϕ values.

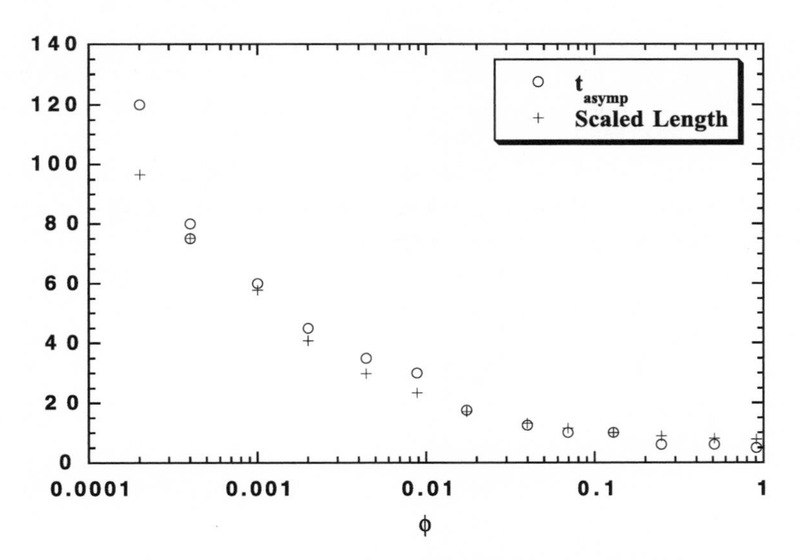

Figure 6.12 $t_{\text{asymp}}(\rho = 0.5)$ vs. ϕ for temporary-removal dynamics, compared with a scaled version of $L(\phi)$ for an equivalent β-graph ($n = 1,000$, $k = 10$).

analogy to the characteristic length of the underlying graph. It might be that a single attractor exists for any one value of ρ, regardless of ϕ, but that the transient time to that attractor depends on the structural characteristics of the graph in a particularly simple manner.

6.3 MAIN POINTS IN REVIEW

For the very simple dynamics considered here, there are already many possible relationships between structure and dynamics. Nevertheless, the following general conclusions appear valid:

1. For a broad range of dynamical parameters, where the system approaches either a steady state or a state in which the partial fractions are asymptotically steady, either: (a) The nature of the attractor is determined by the coupling topology, or (b) The time taken for the systems with different coupling topologies to reach the *same* attractor (characteristic transient time) is determined by the coupling topology. Specifically, spreading occurs faster in systems with shorter characteristic path length, just as one might expect intuitively.
2. No generally applicable relationship between structure and dynamics seems evident, but in a few specific cases (such as the characteristic transient time for permanent-removal systems, in which the diseases takes over the entire population) there exist relatively simple interpretations.

Whether or not the same can be said for systems whose dynamics is less obviously related to their connectivity (i.e., not spreading problems) is the subject of the next three chapters.

7

Global Computation in Cellular Automata

Stringing together phenomena as diverse as chaotic attractors, artificial life, and universal computation, cellular automata (CAs) are a nontrivial and scientifically interesting test bed in which to examine the interrelationship of structure and dynamics. Simple though these constructions are, they still manage to display extraordinarily complex and, at times, fascinating behaviour. This chapter explores only a narrow subspace of this burgeoning field of research, but one that promises much in the way of understanding the architecture and mechanisms of spatially extended, locally connected systems that perform what is known as *global computation*. The background review is correspondingly biased and brushes over or omits much of general interest. For a comprehensive review of the early history of CAs, see Burks (1970), and for more recent developments see Wolfram (1986) and Mitchell (1996a).

7.1 BACKGROUND

The study of cellular automata is a relatively new field that appeared in the midst of the intellectual dust cloud that accompanied the development of the first digital computers. The lion's share of credit for the birth of both ideas can be attributed to just one man: John von Neumann. Whilst his technological brainchild may seem by far the greater innovation, the future of computation in massively parallel systems may yet prove cellular automata to be the more far-reaching insight. Von Neumann, however, was not actually trying to build CAs for computational purposes. Rather, he was answering the question, motivated by the phenomenon of biological reproduction, What kind of logical organisation is sufficient for an automaton to be able to reproduce itself? His solution, called the *self-reproducing automaton* (von Neumann 1966), was a two-dimensional infinite grid of cells on which any automaton (that is, a finite-sized block of cells with prespecified cell states) could be constructed, given information fed to it via a "tape" (an external one-dimensional array of cells containing the coordinates and cell states of the automaton to be constructed). Construction proceeded via a "construction arm" that in reality, was a propagating sequence of states moving through the cells between the "construction control" and the "construction site" where the new automaton was to be located. All these high-level instructions were coded into low-level cellular automata language with the use of twenty-nine cell states from which basic logical operations could be synthesised. It turned out that von Neumann's "machine" was capable of *universal construction* in the sense that it could build any automaton that could be prescribed on the input tape. Thus, not only did self-reproduction follow as a special case, but the automaton could replicate the behaviour of the most general of computational devices: the Universal Turing Machine. It should be noted that von Neumann's machine predated the discovery of the mechanisms by which DNA replicates itself: a truly remarkable case of life imitating art imitating life!

Since those heady days of the 1960s, much work has been done both to refine von Neumann's original ideas in the related worlds of self-reproduction and universal computation and to integrate the concepts of cellular automata with those of dynamical systems theory. Interesting though the history of the former is, it is the latter approach from which the current work springs. It also seems that this is a more fruitful approach because it seeks to emulate the computational features of *biological* systems, thus leaping ahead of, rather than chasing, Turing's shadow.

But what precisely *are* cellular automata? Different researchers have used a number of variations upon the general idea depending on the phenomena they wished to emphasise or emulate. The construction used here (see Fig. 7.1) is due to Stephen Wolfram (1983), who defines a cellular automaton to be a regular lattice of n discrete cells in some dimension, where each cell $i = 1, 2, \ldots, n$ is attributed, at each point in time, a *state* (s_i). This state may occupy any one of a finite number (η) of discrete values and is updated at discrete intervals in time, according to a deterministic *transition rule* (Φ), which depends only on the current state of the cell i and k other cells that are in its immediate neighbourhood $\Gamma(i)$.[1] Hence the update rule

$$s_i(t+1) = \Phi\big(s_i(t), \mathbf{s}_{\Gamma(i)}(t)\big) \tag{7.1}$$

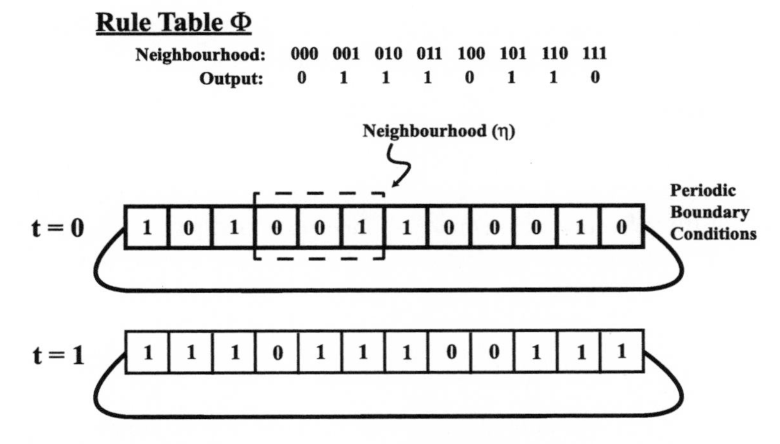

Figure 7.1 A one-dimensional, two-state cellular automaton with one possible rule table for $k = 2$.

is *local* both in time and space. Implicit in Equation 7.1 are the additional simplifications that all cells are identical (in the sense that they all occupy the same range of states and all have the same update rule) and that they are all updated synchronously. Finally, the finite and discrete nature of the CA admits only a finite (though possibly extremely large) range of possible Φ. That is, if each cell must occupy one of only η states and Φ may only depend on $(k + 1)$ of these states (the cell plus its k neighbours), then the number of possible states upon which Φ may operate is η^{k+1}. For each of these states Φ can yield one of only η values, hence the total number of possible Φ is limited to $\eta^{\eta^{k+1}}$. Although the definition admits lattices of any dimension, almost all the work has focused on one and two dimensions, with by far the most attention devoted to one-dimensional, two-state ($s_i \in \{0, 1\}$) automata with k not much greater than two. In this parameter range, where η^{k+1} is not too large, it is frequently convenient to represent Φ as a *rule table,* which lists neighbourhood states and their corresponding cell states explicitly, instead of as a higher-level functional statement. A further representational convenience is possible for one-dimensional automata—a *space-time diagram*—in which successive temporal states are placed beneath each other, and otherwise obscure patterns can be observed in the resulting two-dimensional grid.

When $k = 2$, the resulting one-dimensional, two-state CA is termed an *elementary* CA. Wolfram (1983) studied and classified all 256 possible rule tables of elementary CAs and then later generalised this (Wolfram

1984) to suggest that all one-dimensional CAs fall into one of four universal "classes," the first three of which are analogous to fixed points, limit cycles, and chaotic attractors of continuous dynamical systems, respectively, and the fourth of which seems to be capable of universal computation (in one dimension less than von Neumann). Once again, all this work is deep, intriguing, and original but still tangential to the thrust of the present work, which is concerned mostly with how a *locally connected* structure can perform computations that are, in some sense, *global*.

7.1.1 Global Computation

Work of this sort had actually begun as early as von Neumann's original automaton, in the form of the "firing squad synchronisation problem" (see Mazoyer 1987 for an overview of this problem). A one-dimensional array of cells is imagined to be a squad of soldiers in a line, the leftmost of which is the "commander." At $t = 0$, the commander, responding to an external signal, gives the order to fire. The problem is to find an algorithm (that is, an automaton) such that after some number of time steps have elapsed, all cells turn on (fire) simultaneously. This would be a trivial task for any system that possessed a central processor, but the challenge is that only local communication (that is, from each cell to its immediate neighbours) is allowed. This problem was originally solved with the use of a high-level *particle* description of the process. That is, the commander sends out signals that propagate at certain velocities (defined as the number of cells moved in a given number of time steps, or the slope of the signal in the space-time diagram) and interact with each other and the array boundaries in predetermined fashions. Having developed the appropriate signals and rules, this high-level description can then be converted into the language of basic CA operations on cell states. Unfortunately this approach does not generalise well, as some tasks are too difficult for anyone to develop even the high-level solution, and even if such solutions could be constructed, their low-level implementation might require more cell states or larger neighbourhoods than a particular CA possesses. It therefore seems advantageous to automate the process of finding optimal solutions to problems of global computation using some kind of device that operates directly on the rule tables, and then perform the inverse problem of inferring the high-level rules from the resulting low-level solution. It is this approach that has been proposed and advanced by Mitchell and Crutchfield (Mitchell et al. 1993,

1994, 1997) in the last few years in the context of one-dimensional, two-state CAs.

The Mitchell-Crutchfield approach is based on a two-stage process. First, a population of many randomly generated CA rule tables of fixed size (usually with $n = 149$ and $k = 6$) are set a given task of a global nature (that is, one that would be trivial for a central processor but nontrivial for a locally connected, distributed system). The population is then allowed to "evolve" under the action of a *genetic algorithm* (GA) (described below) until an "optimal" solution is found.

Second, the modus operandi of the best solution is reconstructed from its space-time behaviour (rather than its rule table) by identifying *regular domains* in the space-time diagram and filtering these out. The CA is then represented solely in terms of its domain boundaries, which can be interpreted as *particles*: carriers of information whose interactions form the basis of global information processing.

Both these steps are interesting and difficult problems in their own right and utilise quite independent ideas. The first invokes the use of a genetic algorithm—a device inspired by the process of genetic diversity and evolution in biological systems (see Mitchell 1996b for a comprehensive introduction). In this way of thinking about CAs, the rule table (which is effectively nothing more than a bit string of length 2^{k+1}) is viewed as the *genotype* (or *chromosome*) of the CA, whilst its space-time behaviour (and particularly, its final state after a set number of iterations) is thought of as its *phenotype*, on which selective forces can be applied. In a computational context, the selection criterion is the success rate at which the phenotype of any given chromosome solves a particular problem. The first case studied by Mitchell et al. was the problem of *density classification* (Mitchell et al. 1994): find a CA such that if a particular initial condition (the initial state of the CA) has fewer than half the cells in their *on* state (that is, *density* $\rho_0 < \rho_c = 1/2$), then every cell must eventually turn *off*; conversely, if $\rho_0 > \rho_c$ then all cells must turn on. The point here is that the set of possible solutions (chromosomes) is far too large (2^{128}) for any kind of systematic optimisation search to be feasible, and so another method is required to generate successively better solutions from some initial, randomly selected (and relatively small) population.

Each of the CAs in the sample population is run on a large number (I) of initial conditions, which are selected randomly but in such a fashion that ρ_0 is uniformly distributed over the interval $[0, 1]$. The *performance fitness* (F) of each chromosome is computed as the fraction of correct classifications made by its CA (after $2n$ time steps) over the set of I initial conditions. The chromosome population is then ranked in descending order of F and a number of them selected, without modification, to survive into the next generation. The remaining members of the population are replaced by new members that are formed by genetic crossover and mutation of the bit strings of the successful chromosomes. This process is repeated for each generation until the optimal performance fitness of successive generations plateaus. Once an optimal strategy has emerged in this fashion, its *unbiased performance* P is computed using initial conditions whose individual states are set independently (yielding a *binomial* distribution in ρ_0 that is peaked around the "hard" case of $\rho_0 = 0.5$).

The second stage concerns the analysis of the most successful CA and so requires an ability to extract patterns out of arbitrarily complex structures in some consistent, automated fashion. The basic idea, as introduced by Crutchfield (1994), builds upon Wolfram's classification of elementary CAs by quantifying their *computational capacity* in terms of the highest level of regular language needed to yield a *finite description* of the space-time evolution of an automaton. Basically what this boils down to is a search for regular, repeating patterns (for instance, every other site is a 0) and then filtering the patterns out of the space-time diagram of the CA. All that remains after this process is the cells that do not conform to the patterns and so form boundaries between the regions of cells that do. These boundary cells are referred to as *embedded particles*, and they can then be examined to see if their space-time behaviour obeys some definable (that is, finite) rule, such as "the embedded particles undergo random walks and annihilate when they collide with one another." If such a rule does exist (and can be found), then the CA has been explained and its computational capacity determined to lie at that level (that is, the particle level). If no such rule exists, then the process must be repeated, once again trying to extract patterns out of the mess of *particles* this time, yielding perhaps *meta-particles* and so on. The level of structure required to describe the CA's behaviour completely defines its *intrinsic computation*, which Crutchfield argues places an upper bound on the complexity of any task that it can be expected to perform.

All this is relevant to the task of understanding global computation in CAs (developed by genetic algorithms) because it provides a general, systematic method of moving from the impenetrable low-level rules embodied in the chromosome to high-level rules that illuminate the crossover from local to global computation. This work shows considerable promise for understanding the ability of one-dimensional, locally connected structures to perform tasks of global computation. However, it is not easy to see how the approach could be generalised to higher dimensions, and certainly the systems of which CAs are supposed to be heuristic representations are not one-dimensional lattices. Thus, in a context of network structure, it is natural to ask whether or not high-performance CAs can be developed *by varying the coupling topology of the automata instead of their rules*.

7.2 CELLULAR AUTOMATA ON GRAPHS

A specific way to approach this quite general question is to consider the two CA tasks that have already been thoroughly explored with the genetic algorithm/particle approach and recast them, using rules that Mitchell et al. have shown to be *unsuccessful* in one-dimensional, but situating them instead on the graphs constructed in Part I. The two tasks at hand are (1) the *density classification problem*, in which a CA is expected to turn all its cells *on* (after some finite time) if more than half its cells are on in the random initial condition, and *off* otherwise, and (2) the *synchronisation* problem, in which a CA is expected (again in some finite but unspecified time) to turn all its cells on and off at alternating time steps, regardless of its initial condition.

7.2.1 Density Classification

The density classification problem is simple from a global perspective, but not so simple from the local perspective of a vertex in a one-dimensional lattice. That is, the obvious way to determine whether less or more than half of the initial states are on is just to count them. But to do that, one requires information about *at least half* of the n cells in the entire system. However, each cell in the CA is connected only to k other cells, where it is specified that $k \ll n$. Hence, although it might be easy for a central processor to do the job (just run through all the cells and count how many are on), no individual cell is capable of doing this. And this is the problem: a *global* computation is required, but the system is only *locally* connected. In fact, Mitchell et al. show that the simple-minded intuition of the CPU approach applied locally (each cell sorts through its neighbourhood and copies the state that is occupied by the majority of the neighbours) almost never works in a one-dimensional space because the system gets stuck in a mixed state of multiple all-on or all-off *domains* (Mitchell et al. 1994). Within each domain, each cell is adjacent to a majority of like-minded cells and so will never change. Meanwhile, the cells at the domain boundaries just stare at each other, unmoved. Unless, by sheer good fortune, the initial condition is such that one domain can fill up the entire system (say, almost all cells are on or off to start with), the system will always get stuck. Consequently, the performance of what they call *majority rule* is dismal.

Mitchell et al. set up majority rule as a strawman for their genetic algorithms to replace, and as an illustration of what sort of things a locally connected "global computer" should *not* do. But on small-world graphs, maybe it's not such a dumb idea after all. That is, even though small-world graphs are still sparsely connected, we already have the sense that they are qualitatively different from straightforward 1-lattices. Indeed, one of the conclusions of Chapter 4 is that in small-world graphs, the local and global length scales coincide, suggesting that the distinction between locally and globally connected may be, in their case, misleading. So perhaps (the intuition goes) the cells of a CA constructed on a small-world graph will have sufficient information about the rest of the world to act like CPUs and get away with it.

Because, as ϕ increases from 0, the degree k has nonzero variance (that is, not every cell will get the same number of neighbours), the term *majority* can be ambiguous. Nevertheless, an appropriate variation of the majority-rules CA can be defined as follows:

1. Generate an initial condition over n cells, where each cell has equal probability of being on or off.[2]
2. For each time step $t > t_0$, each cell v counts the number of on cells $k_{\text{on}}(v)$ in its neighbourhood $\Gamma(v)$. It then alters its state according to the following conditions:

   a. If $k_{\text{on}}(v) > k_v/2$, then turn on.
   b. If $k_{\text{on}}(v) < k_v/2$, then turn off.
   c. If $k_{\text{on}}(v) = k_v/2$, then decide on or off randomly, with equal probability.

3. Repeat until a steady state is reached or maximum allotted time ($t_{\max} = 2n$) expires, at which point the final fraction of on cells ρ_f is recorded.
4. If ($\rho_0 > 0.5$ and $\rho_f = 1$) or ($\rho_0 < 0.5$ and $\rho_f = 0$), then the CA has correctly classified the initial density.

5. After 100 independent realisations, compute the unbiased performance P.

Before actually testing this out, it should be stressed that the result is not obvious either way.[3] Perhaps in a fully random world it might seem intuitive that, because any local neighbourhood will be a random sample of the entire system, then a local majority rule will work. Indeed, political pollsters rely on the representativeness of random samples every day. But it is not obvious that this approach will work for such a precise task as density classification, and it is certainly not obvious that it will work on a small-world graph, which has far less than all its edges allocated randomly. Also, there is no obvious "spreading" dynamics in this system, like there was in the previous chapter. We can think vaguely of *information* spreading between neighbourhoods, but this certainly doesn't work in the one-dimensional case and so won't necessarily work for the small-world case either. Finally, note that the unbiased performance P is not as obviously related to coupling topology as $S(t_{\max})$ (uninfected fraction of the population) and t_{steady} (the time for the disease either to die out or take over the entire population) were for disease spreading. Rather, P is a measure of how often the CA "gets it right," where a correct answer depends on both the initial and final states of the system. It is certainly not obvious that such a high-level statistic must necessarily have any simple relationship to the coupling topology at all.

Nevertheless, Figure 7.2 shows that this is indeed the case. For the parameters used by Mitchell et al. ($n = 149$, $k = 6$) and $\phi = 0$, $P \approx 0$, as one would expect for a 1-lattice, for the reasons given above. However, for even a small increase in ϕ beyond 0, rapid and substantial gains in P are made, levelling out eventually at $P \approx 0.9$. This is particularly impressive, given that the most successful GA-generated solution had $P = 0.769$ and the best human-generated rule in 1-lattices (the GKL rule) yields $P = 0.816$ (Das et al. 1994). For larger values of n, the difference is even more striking. Figure 7.3 shows P versus ϕ for $n = 146$, $n = 599$, and $n = 999$, and Table 7.1 compares the maximal values obtained on small-world graphs with the other methods mentioned above.

From these results, it appears that the small-world graph approach both outperforms the other available methods at parameter values for which data are available and is less sensitive to increasing n, implying that the relative performance advantage would increase with respect to n. In fact, it appears from Figure 7.3 that $P(\phi)$ converges to some limiting function as n increases. This, bear in mind, is for $k = 6$, which does not

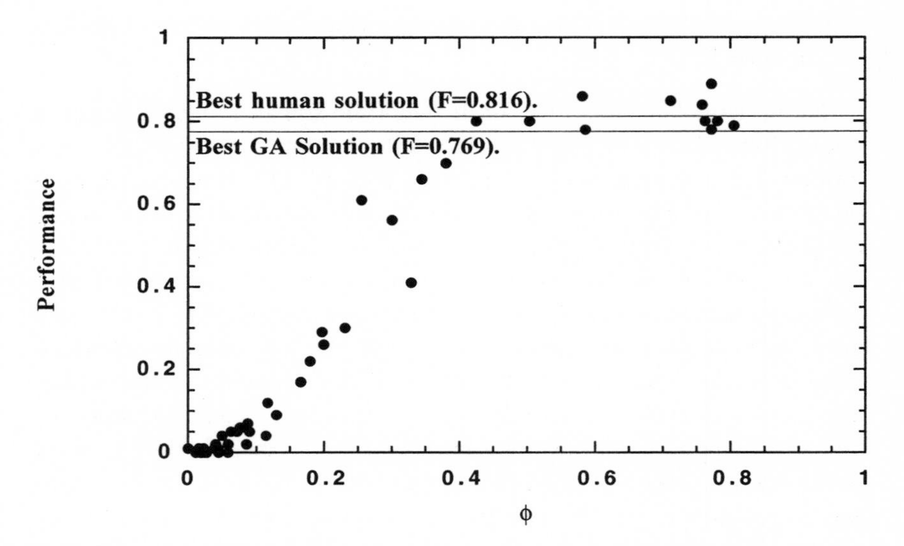

Figure 7.2 Unbiased performance (P) vs. ϕ for the density classification task, where solutions are generated by the majority-rules strategy operating on β-graphs ($n = 149$, $k = 6$). The results are compared with those of the best human-generated and genetic-algorithm-generated strategies for 1-lattices (Das et al. 1994).

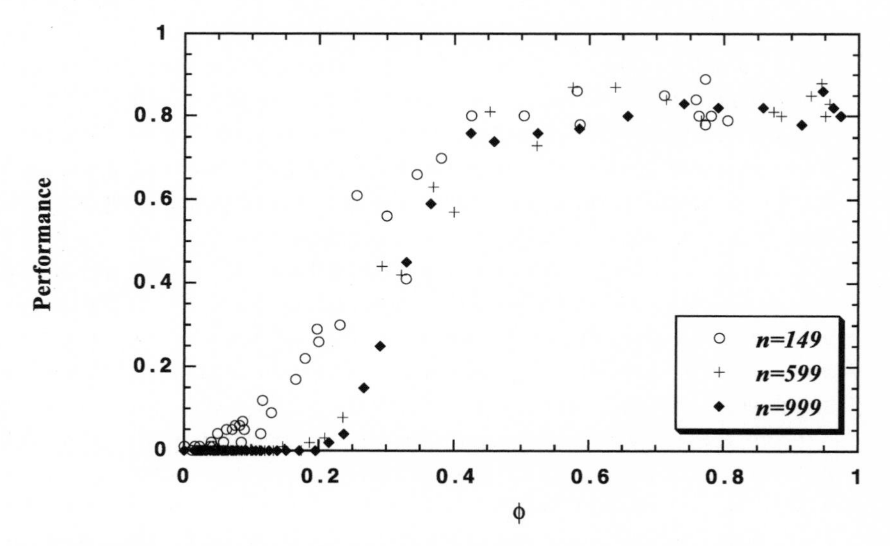

Figure 7.3 Comparison of unbiased performance (P) vs. ϕ for the density classification task on β-graphs for $k = 6$, $n = 149$, $n = 599$, and $n = 999$.

TABLE 7.1
Unbiased Performance of β-graphs Compared with the Best Genetic Algorithm and Human-Generated Results

n	GA	GKL	Graph ($k = 6$)
149	0.769	0.816	0.89
599	0.725	0.766	0.88
999	0.417	0.757	0.86

really satisfy the condition $k \gg 1$. However, the same tests as above can be run for $k = 12$ to see if any improvement in performance results. Figure 7.4 shows that the maximal performance attained improves slightly for all n, and Figure 7.5 shows clearly that the performance "cliff" for $n = 999$ occurs at smaller ϕ for $k = 12$ than it does for $k = 6$.

One feature of all these results deserves special mention: the performance cliff occurs well before the random graph limit is reached but well after the corresponding *length* transition for the underlying graph (see Fig. 7.6). This is another perplexing scenario, in which it appears that the dynamics can be dramatically influenced by only subtle changes in coupling topology *and* in which some small-world topologies appear to exhibit some of the dynamical features of a random topology, yet it is hard to see exactly how this comes about.

Intuitively, one might expect that more shortcuts might be required for high performance than for a low characteristic length because a single shortcut can contract the path length between many pairs of vertices, but, as far as the CA is concerned, it provides an information link between only two neighbourhoods. In fact, it seems plausible that *every* cell would require its own shortcuts before it could be capable of sampling the population in a representative sense. This reasoning would lead one to anticipate a critical value of ϕ (near which dramatic changes in P would occur) at approximately the value where each vertex would be expected to possess one shortcut edge. That is, if $nk/2$ is the total number of edges in the graph, then $\phi nk/2$ is the total number of shortcuts and $\phi k/2$ would be the expected number of shortcuts per vertex. If the critical value of ϕ (ϕ_{crit}) is such that, on average, each vertex would have one shortcut, then $\phi_{\text{crit}}(k/2) \approx 1$, yielding

$$\phi_{\text{crit}} \approx \frac{2}{k}. \tag{7.2}$$

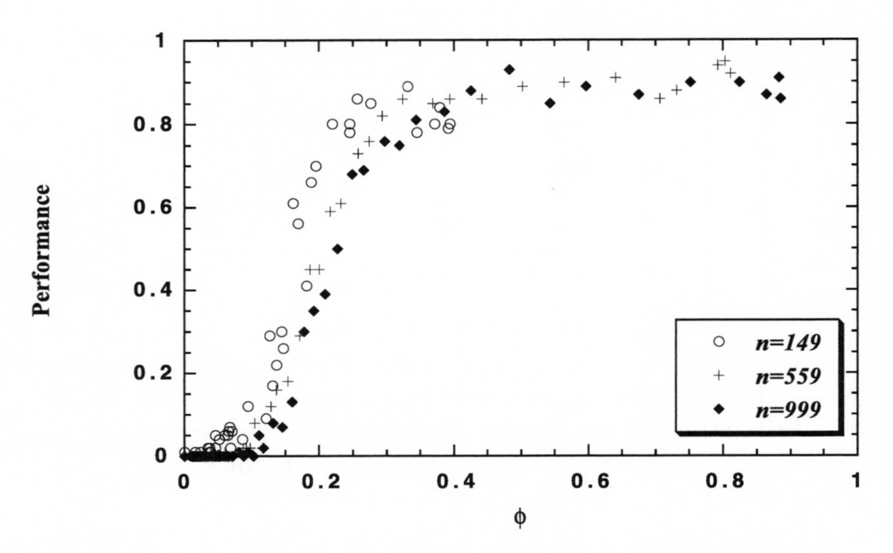

Figure 7.4 Comparison of unbiased performance (P) vs. ϕ for the density classification task on β-graphs for $k = 12$, $n = 149$, $n = 599$, and $n = 999$.

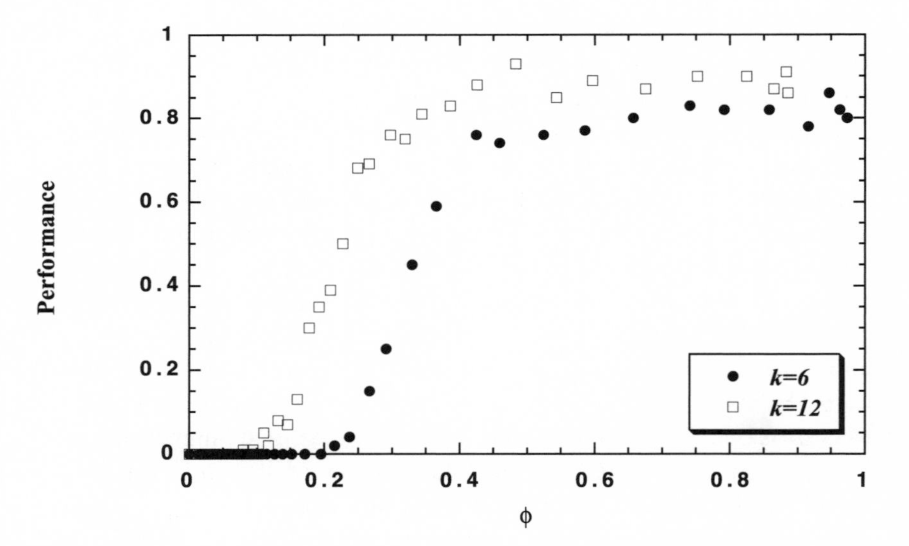

Figure 7.5 Comparison of unbiased performance (P) vs. ϕ for the density classification task on β-graphs for $n = 999$ and $k = 6$, $k = 12$.

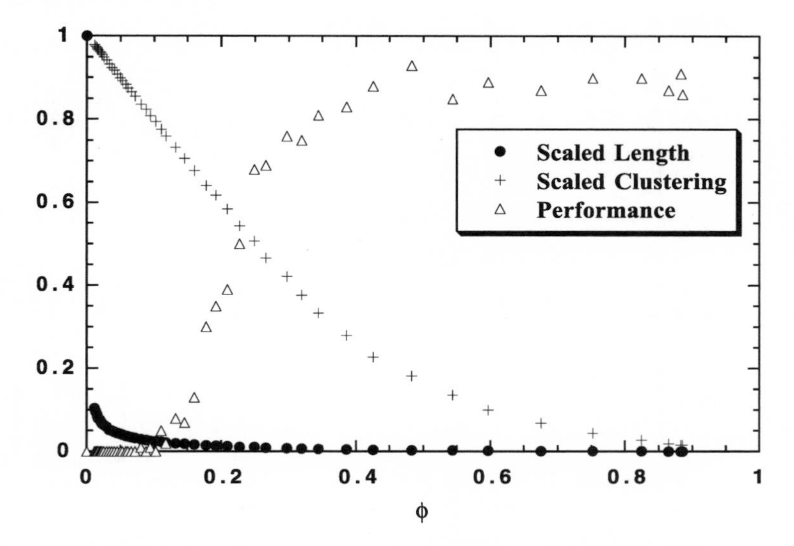

Figure 7.6 Functional comparison of characteristic length (scaled for comparison), clustering coefficient, and performance fitness for a density-classification CA on a β-graph ($n = 999$, $k = 12$).

Judging by Figures 7.3–7.5, this estimate seems on target. Figure 7.5 is particularly instructive in this respect, as it indicates that doubling k causes ϕ_{crit} to be reduced by approximately the same factor ($\phi_{\text{crit}} \approx 0.1$ for $k = 12$, down from $\phi_{\text{crit}} \approx 0.2$ for $k = 6$), just what one would expect from Equation 7.2.

As a check on the robustness of these results, it is worth demonstrating that they are not dependent upon the choice of model used to obtain them. Figure 7.7 shows a comparison of results for $n = 999$ and $k = 12$ where both α- and β-models have been used. The results are indistinguishable, which is interesting, as ϕ was motivated by a desire to unify the *length* properties of the α- and β-models and is not obviously the correct statistic for unifying the performance fitness of the corresponding CAs. The fact that it does lends support to the claim that graph structure (as measured by ϕ) has a strong and consistent effect upon the dynamical properties of the system. In the next chapter, however, it turns out that this claim does not hold so well for other dynamical systems, so it shouldn't be emphasised too much here.

Before leaving the density-classification problem, it is interesting to compare the performance of majority-rules CAs constructed on spatial graphs with those on relational graphs. Figures 7.8 and 7.9 both show,

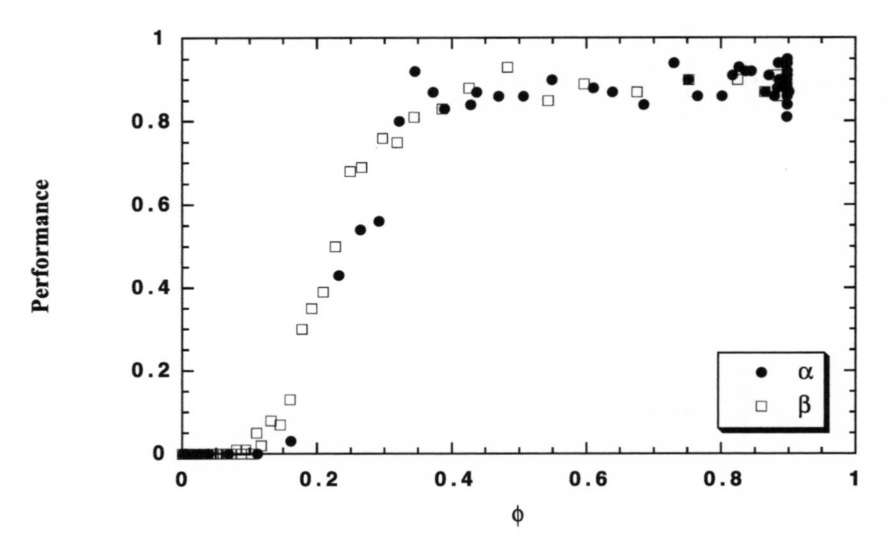

Figure 7.7 Comparison of unbiased performance (P) vs. ϕ for the density classification task on relational graphs constructed using α- and β-models ($n = 999, k = 12$).

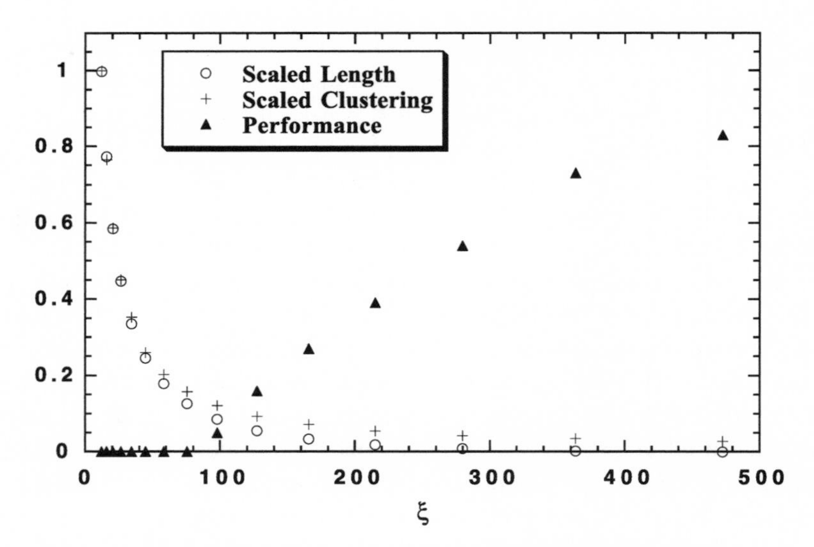

Figure 7.8 Functional comparison of characteristic length (scaled for comparison), clustering coefficient, and performance fitness vs. ξ for a density-classification CA on uniform spatial graphs ($n = 999, k = 12$).

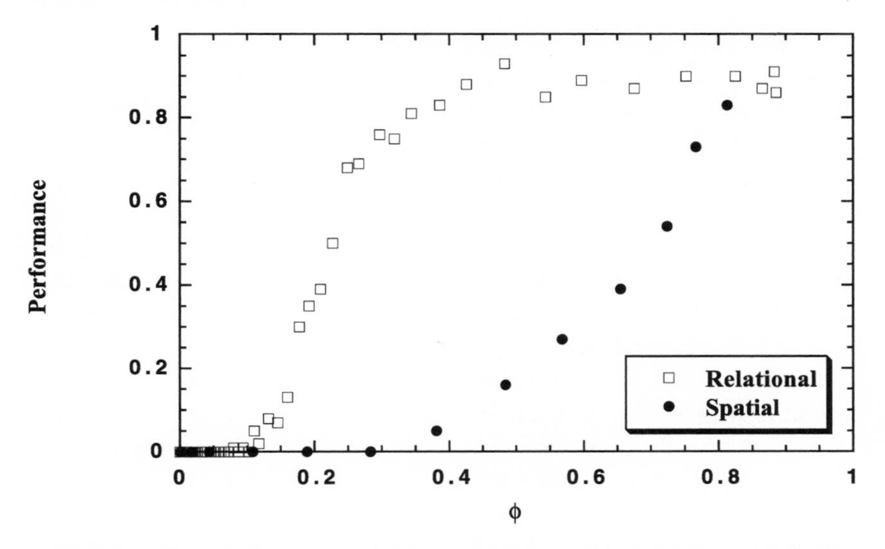

Figure 7.9 Comparison of unbiased performance vs. ϕ for spatial and relational graphs ($n = 999$, $k = 12$).

in different ways, that spatial graphs support only optimal performance in the random graph limit. Figure 7.8 indicates that high-performance fitness is achieved only once the clustering coefficient γ has approached its random-graph limit. Figure 7.9 supports this conclusion, comparing the spatial and relational-model performance for the same parameters. From these results, it seems clear that relational graphs (which exhibit small-world features) significantly outperform spatial graphs (which do not), indicating that small-world graphs constitute an interesting class of objects from a *computational* as well as a structural perspective.

7.2.2 Synchronisation

The second test case considered here was suggested and examined by Das, Mitchell, and Crutchfield (1995) and is a variant upon the same theme as the density-classification problem. In the *synchronisation problem*, the task on which the CA is assessed is its ability to end up in a globally synchronised, alternating state (all-on followed by all-off) regardless of the initial condition. On the surface of things this seems like a qualitatively different problem from density classification, and indeed Das et al. followed a similar but independent procedure to that which they used in the density classification problem in order to develop successful rules for the synchronisation task.

It turns out, however, that a simple variant of majority rules works just as well for synchronisation as it does for density classification. In fact, when recast in this fashion, they turn out to be almost exactly the same problem. Specifically, the *contrarian* rule operates precisely the same as majority rules except that when a majority of a cell's neighbourhood is on, the contrarian rule turns the cell off and vice versa. That is, it does precisely the *opposite* of the majority. This is an obvious candidate for the synchronisation problem as, once a group of cells are in the same state, they will alternate synchronously to opposite state each successive turn. It is not immediately obvious how they get to a synchronised state in the first place, but some simple reasoning can provide at least the intuition. If two CAs are started off with identical initial conditions, one with majority rules and the other with contrarian, then after a single time step they will be in opposite states, regardless of their connectivity. This is because, in each case, every cell looks at its neighbourhood and makes its decision based on the same information and at the same time as every other cell. But the corresponding cells in the two CAs make opposite decisions, so the CAs end up in opposite states. After another time step, the same reasoning demonstrates that they will occupy the *same* state, then the *opposite* state, and so on.[4] Hence if the CA operating according to majority rules synchronises eventually, then so must the contrarian CA, at which point it must necessarily alternate its entire state synchronously as required by the task. The only question, then, is whether or not the majority rule will synchronise. But we already know the answer to this from the analysis of the density classification problem: it depends on the coupling.

Figure 7.10 shows a representative comparison of results for the two tasks, using the appropriate rules on relational graphs of varying ϕ, for $n = 149$, $k = 6$. Two features are immediately obvious: (1) the functional forms of $P(\phi)$ for the two problems are virtually identical, and (2) the maximal fitness of the synchronisation problem is greater.

The explanation for the first of these observations is given above: they are virtually identical tasks. The explanation for the second arises from the fact that, occasionally, the classification CA will lock into the *wrong* state (that is, turn all on/off when the initial condition was less/more than half on), but the synchronisation CA synchronises anyway and so still scores. Hence the unbiased performance of the synchronisation CA is just

$$P_{\text{synch}} = 1 - \mathrm{P}[\text{CA gets stuck in a domain state}].$$

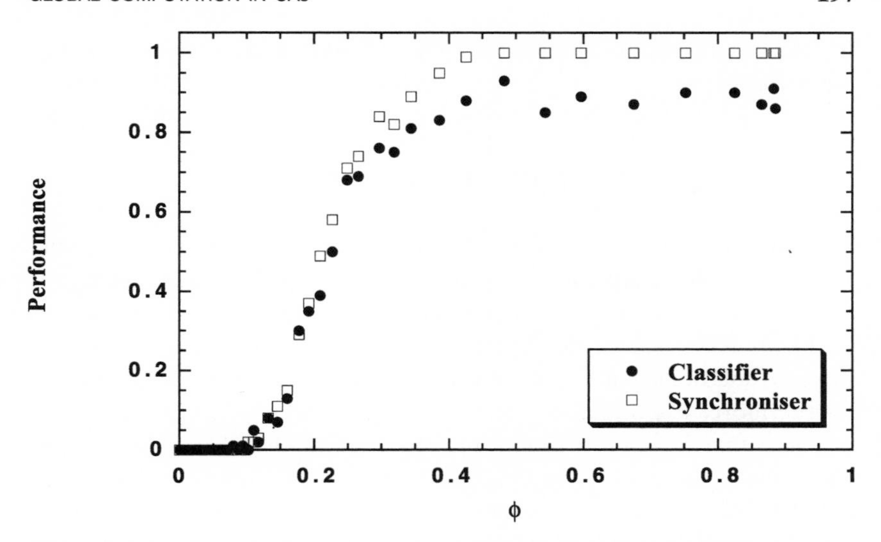

Figure 7.10 Comparison of unbiased performance between the majority-rules CA (operating on the density classification task) and the contrarian CA (operating on the synchronisation task) for $n = 999$, $k = 12$.

For $\phi = 0$ this almost always happens, so $P \approx 0$, and for large ϕ, almost never, so $P \approx 1$. This, incidentally, is about the same as the performance of the best GA-generated rule found by Das et al.

From the two examples considered here it might be tempting to draw the conclusion that a small-world cellular automaton can function as an approximation of a mean-field CA (a CA in which every cell is connected directly to every other). Certainly this appears to be the case with the classification and synchronisation problems in which the rules basically assumed that each cell had some representative knowledge of all others. But this is naive and will not work for more complicated tasks in which it *is* necessary to have knowledge of *every* cell or where it may not be obvious even how a mean-field-type solution should be formulated. An example of the first class of problem is the firing squad synchronisation problem, mentioned earlier, where a single cell is identified as special and must transmit its information to all others. In this case, it is not obvious how a small-world CA should be designed (if indeed it could be made more efficient than a one-dimensional solution), but certainly the mean-field analogy would be invalid. An example of the second class of problem would be any complex, multifaceted task (like winning a game of chess) where it just isn't obvious how *any* CA would approach a solution. In both these cases then, an automated approach, perhaps using genetic algorithms that operate *on both rules and connectivity*, seems advisable.

7.3 MAIN POINTS IN REVIEW

This chapter suggests an alternative approach to solving global computational problems with locally connected systems, by manipulating the *architecture* of the system rather than its rule base. At least for the two problems investigated, this approach seems to work at least as well as the best solutions generated with the currently-prevailing GA approach. Furthermore, it generalises easily to larger n and k—an aspect of the problem that has posed significant obstacles to the GA approach (because the rules have to be developed on CAs of fixed size). Naturally, if the problem at hand demands a strict one-dimensional architecture, then this kind of meddling is not permitted, and the GA approach seems more attractive. But perhaps this is the point: *natural* computational systems are *not* strictly one-dimensional architectures. In fact, Chapter 5 suggests that the connectivity of many natural networks is better represented by small-world graphs than by many other plausible models, including one-dimensional architectures. Hence elegant and attractive though the one-dimensional GA/particle approach may be, the broader project of understanding the computational capabilities of *real*, locally connected computational systems should probably account for the dramatic effects of small-world architectural topologies in which the traditional two-dimensional space-time diagrams cease to be meaningful. Perhaps the most fruitful approach to solving these kinds of problems might be the application of genetic algorithms to the combined space of all possible rules *and* all possible connectivities.

8

Cooperation in a Small World: Games on Graphs

The theory of games is a natural next step to make after cellular automata in considering dynamical systems and their behaviour as a function of their coupling topology. Inasmuch as multiplayer, iterated games are both temporally and spatially discrete and exist in a finite state space, they share many features of CAs. But their governing rules comprise a special case of the complete set of CA rules that are of particular interest—namely, rules that can be interpreted as cooperation with or exploitation of other players. In this sense, games may be simpler than the more general CAs. However, they are also more complex in that it is natural for them to violate the condition of homogeneity—people do not all play by the same rules—and it is of considerable interest to social and biological scientists alike which rules, in a population of *interacting* players, will prevail. The emphasis in game theory is thus shifted away from the inherent computational capacity of the system, where all elements operate in concert to achieve a single goal, to *competition* between elements defined by some externally imposed measure of *individual* performance. These additional complexities lead naturally to a different set of questions than those posed for cellular automata: (1) For a homogeneous population in which every element is capable both of cooperation and exploitation, how does the coupling topology of the system affect the *emergence* of cooperation? (2) For a heterogeneous population evolving over a series of generations, and in which successful rules are passed on preferentially to future generations, how does the coupling topology affect the *evolution* of cooperation? First, however, some background.

8.1 BACKGROUND

John von Neumann was a remarkable man. In the optimistic aftermath of World War II, at the dawn of the American hegemony of the sciences, von Neumann was a key player who seemingly by himself invented whole fields of research. At about the same time as he was designing the first digital computers and carving out the brand new theory of cellular automata (see Chapter 7), he also produced what has since been recognised as the seminal work in the theory of games.[1] In his *Theory of Games and Economic Behaviour* (von Newmann and Morgenstern 1944) von Neumann emphasised economic applications of game theory, focusing on the optimisation of a rational player's *utility function* in zero-sum games, usually involving only a few players. This was the traditional economists' view of things (Oskar Morgenstern, von Neumann's co-author, was an economist): that for someone to win, someone else had to lose and that, in a world of perfect information, optimal strategies could always be formulated (either deterministically or stochastically) on the basis of purely rational behaviour. John Nash developed von Neumann's work further, proving the existence of equilibria (optimal strategies) in zero-sum games with an arbitrary number of players (Nash 1951) and also for two-player, non-zero-sum games in which, through bargaining, both players could stand to benefit (Nash 1950, 1953). Nash was awarded the 1994 Nobel Prize in Economics for these contributions, which became the analytic foundation on which the economic applications of game theory are constructed.

Psychologists, however, were interested in a different kind of game in which players had a mixture of competing and common interests, but in which economic-type bargaining was not permitted. The surprising result of this restriction was that the action that, on the surface of things, should maximise a player's expected payoff could no longer be counted on to produce the optimal result, even given perfect information. This paradox is resolved by the observation that by cooperating and accepting small penalties, all players receive higher payoffs than if they adopt rational positions predicated on an attitude of pure conflict. Simply realising this, however, does not reduce the problem to the prescriptive notions of zero-sum games. Hence exactly how much to cooperate and how this depends on the parameters of the game are problems that remain, in general, unresolved. Although various models were proposed to capture the essence of this conundrum, by the mid-1960s a single model, known as the Prisoner's Dilemma, had emerged as the archetype of this class of non-zero-sum games.

8.1.1 The Prisoner's Dilemma

The Prisoner's Dilemma is a mathematical abstraction of the following situation. Two prisoners, accused of the same crime, are being held in separate cells and are both offered the same deal by the police: sell the other out and receive a lesser punishment. If one cuts a deal and the other remains faithful to their compatriot, the sneak gets a reprieve and the naive friend gets slammed. If they both rat on each other, then they both get punished but not as badly as if only one had borne all the responsibility. Finally, if they both stay silent they will each receive a lesser punishment still, because the authorities can't really pin anything on them but will still get them for something trivial. The dilemma arises because, from either prisoner's perspective, the best action is always to defect. If one partner defects, it is better to rat on them too than to get suckered. And if they keep quiet, it is *still* better to defect (and get a reprieve) than to cooperate (and get even a trivial punishment). So either way, it is always better to defect. But both players are rational, and both are in the same situation with the same information. So *both* defect. The payoff for mutual defection (which is the second-worst of all possible results) is not optimal for either player. In particular, mutual cooperation (the *irrational* choice) turns out to yield a higher payoff than mutual defection for both players. But if one player *knows* the other will cooperate, the optimal action is still to defect, and likewise if they know their friend is ready to talk. At this point the dilemma should be clear: there is no *rational* way out of the suboptimal solution, and no amount of additional information can change this, bearing in mind that they can't bargain with each other.

A natural generalisation of the dilemma is to have the same players interact many times over, under the same conditions. Given that the players know they will be meeting again and again, it may seem obvious that some kind of cooperation must necessarily emerge. Surprisingly however, the same reasoning that leads to mutual defection for a single game also leads to mutual defection over *any known, finite number of rounds*. If each player knows how many rounds (say, t of them) are to be played, then whatever sequence of choices they might adopt should have defection as its last move. This follows from the fact that the other player will not be able to retaliate to anything done in the last move, and so the optimal, rational thing to do is to exploit that and so defect. Of course, both players being rational, both of them come to the same conclusion. So whatever t is, on the t-th turn both players will defect. This effectively removes the t-th turn from consideration, seeing as it is completely determined regardless of any actions preceding it, which makes the $(t-1)$-th turn effectively the last turn of the game. But here we find the end-game conditions repeating themselves, and once again both players are forced by their own rationality and desire to optimise their utility into a suboptimal choice. This process of "cutting off your nose to spite your face" then spirals all the way back to the first turn, ensuring that two rational, self-interested players will proceed to damage their own and each others' interests by defecting on every single turn.

In mathematical terms this odd situation is represented by the following payoff matrix:

	C_2	D_2
C_1	(R, R)	(S, T)
D_1	(T, S)	(P, P)

where T = temptation to defect, S = sucker's payoff, R = reward for cooperation, and P = punishment for defection. The essential conditions that create the dilemma are (1) $T > R > P > S$, and (2) $(T + S)/2 < R$.

The first of these conditions ensures that a rational player, observing that $T > R$ and $P > S$, will always choose the second row/column, which leads inevitably to the suboptimal (P, P). The second condition is equivalent to the statement that the prisoners cannot escape their dilemma by taking turns exploiting each other. That is, mutual cooperation is superior to alternating T and S. A number of other conditions are also prescribed to prevent any "cheating" in the iterated Prisoner's dilemma:

1. Players cannot communicate or bargain.
2. Players cannot make enforceable threats or commitments.
3. Players cannot know what the other will do on current or future moves.
4. Players cannot eliminate each other or run away from the interaction.
5. Players cannot change each other's payoffs.

Thus, based only on their knowledge of past actions, each player must formulate a strategy to optimise their own payoff in the long run. As pointed out above, in any game of fixed duration, rational players will always defect. Of course, real people do not do this, as Rapoport (of random-biased nets fame; 1965) determined with his experimental studies of the iterated Prisoner's Dilemma:

> Confronted with this paradox, game theoreticians have no answer. Ordinary mortals, however, when playing the Prisoner's Dilemma

> many times in succession hardly ever play *DD* one hundred percent of the time. To be sure, long stretches of *DD* choices occur, but also long stretches of *CC* choices. Evidently the run-of-the-mill players are not strategically sophisticated enough to have figured out that strategy *DD* is the only rationally defensible strategy, and this intellectual shortcoming saves them from losing.

Axelrod (1984) found a rational way out of the paradox, which is that the game needs to be iterated, with significant emphasis placed upon future rounds, but that the *precise number* of rounds remains unknown. Thus each player knows only with a certain probability that they will interact with their opponent again. In this case, which is arguably the most realistic of scenarios so far presented, cooperation does play a significant role, and finally the rational and the reasonable begin to coincide. In fact, a whole range of possible strategies, utilising both cooperation and defection, qualify as rational strategies. The bad news is that, under these conditions, there is no best strategy to be adopted that maximises a player's utility without knowing what strategy the other player is using. Axelrod demonstrated this comprehensively with his now famous computer tournament. Leading game theorists were invited to submit strategies, all of which were subsequently pitted against each other in a round-robin tournament. The victor was Tit-for-Tat, a strategy submitted by Rapoport and in fact one that he observed in his human experiments almost two decades earlier. Perhaps the single greatest appeal of Tit-for-Tat is its simplicity: on the first round Tit-for-Tat always cooperates and then, on subsequent rounds, mimics its opponent's action of the previous round. Hence it embodies the principles that Axelrod later determined were the keys to success under a wide variety of environments, in which future moves are sufficiently important:

1. It is *nice* (it is never first to defect).
2. It is *retaliatory* (as soon as an opponent defects, so does Tit-for-Tat).
3. It is *forgiving* (once an opponent stops defecting, so does Tit-for-Tat).
4. It is *transparent* (it is simple for an opponent to predict Tit-for-Tat's behaviour).

Even after the results of the first tournament were made public and a second, larger tournament organised, not one strategy could *consistently* outperform Tit-for-Tat, so good was it at eliciting cooperation from other strategies and thus (despite falling prey to occasional exploitations) almost always scoring well itself.

Axelrod's work was a major step in the development of game theory and led to a much deeper understanding of how cooperation might evolve in a population of competitive agents in the absence of any Hobbesian central authority and lacking even the traditional evolutionary-biology mechanisms such as kin selection; that is, through *reciprocity*. Axelrod managed to show not only that cooperation through reciprocity (in the form of the Tit-for-Tat strategy) was a reasonably optimal strategy under a wide range of conditions, but also that it satisfied the stronger condition of *evolutionary stability*, which requires that a strategy (1) can thrive in a heterogeneous environment, (2) once fully established, can resist invasion by another single strategy,[2] and (3) can establish itself from relatively humble beginnings in a much larger, non-cooperative population.

This is dramatic stuff, and Tit-for-Tat has since dominated the game theory literature. However, one important aspect of this work is that it largely overlooks the structure of the population. Axelrod did consider the effects of preferential mixing (i.e., cooperators are more likely to interact with other cooperators than with defectors), but in real populations people often have their choices constrained by the network to which they belong. Furthermore, the dilemma considers only interactions between two players, whereas real systems pose issues of cooperation between individuals, within small groups, or even between an individual and the rest of society as a whole. Finally, it is assumed that the only strategies available to players are those prescribed at the beginning of the game. Much work has since been done to come to grips with the removal of these initial simplifications.

8.1.2 Spatial Prisoner's Dilemma

Nowak and May produced a series of papers (see Nowak and May 1992, 1993; Nowak et al. 1994 for the main results) that explored the additional complexities encountered when the iterated Prisoner's Dilemma is played in a two-dimensional grid, on which every player uses the *same strategy* and interacts only with its immediate nearest neighbours. Starting from certain spatial distributions of initial cooperators and defectors, each player plays every other player in its neighbourhood and then chooses its subsequent action by copying the action of the neighbour with the highest score. This strategy is simpler than Tit-for-Tat and emphasises the spatial element of the game. Using mostly computer simulations,

Nowak and May found conditions under which a small seed of initial cooperators could invade a population of initial defectors, and they examined the statistics of the resulting clusters. They also observed complex and beautiful patterns whose spatio-temporal evolution depend quite sensitively on the initial conditions and their single parameter, which can be interpreted as the degree to which the payoff for mutual cooperation exceeds that for mutual defection. Finally, they generalised their results to incorporate effects such as random errors, random vacancies in the lattice, three-dimensional lattices, and (in response to criticism by Huberman and Glance 1993) asynchronous updating of the players' actions: that is, players update their actions at randomly distributed times, rather than all at once, as if driven by a clock.

Another modification to the spatially extended, iterated Prisoner's Dilemma was made by Herz (1994), who analysed the dynamics of the spatial, iterated game, but in which all players utilise Nowak and Sigmund's Win-Stay, Lose-Shift strategy (1993). Win-Stay, Lose-Shift implements a Pavlovian psychology of remaining in its current state (either cooperation or defection) until its payoff is less than some benchmark payoff, at which point it switches to the opposite state. Herz outlined a number of classes of games that could occur for different parameter values with a Win-Stay, Lose-Shift strategy and considered the possibility of different coupling arrangements. These arrangements, however, differ primarily in the number of neighbours rather than the topology of the neighbourhoods, and in each of them explicit global information is contained in the form of the benchmark performance, which is taken to be the expected performance across the entire population.

Finally, Pollock (1989), following work by Boyd and Lorberbaum (1987), considered the iterated Prisoner's Dilemma played on one- and two-dimensional lattices, between a dominant Tit-for-Tat strategy and groups of invading, competing strategies. Pollock showed that the spatial structure of the population helped ensure the evolutionary stability of Tit-for-Tat, whereas Boyd and Lorberbaum (in the absence of spatial structure) had concluded that Tit-for-Tat could always be invaded by an appropriate *mix* of strategies.

It is clear then that the introduction of population structure can have important effects on the outcome of the iterated Prisoner's Dilemma. However, all the work considered above restricts itself to interactions on a one- or two-dimensional lattice and so is not realistic for the majority of situations in which cooperation actually emerges (i.e., social and ecological systems). Also, in none of this work has the emphasis been upon the coupling topology in the sense of comparing otherwise identical games with different network topologies. Recently Cohen, Riolo, and Axelrod (1999) have done precisely this, studying the Prisoner's Dilemma on two-dimensional grids and regular random graphs (with $k = 4$). They find that randomly connected networks are less likely to support cooperation than regular lattices—a result that is corroborated here.

8.1.3 *N*-Player Prisoner's Dilemma

All the work reviewed so far has dealt either with the Prisoner's Dilemma played strictly between two players or else a multiplayer game that consists of multiple, coordinated two-player games. In many ways, this is not a bad approximation in that interactions of individuals in populations are frequently of the multiple, parallel, two-player nature. Boyd and Richerson (1988) consider a slightly more general case in which a population is divided into segregated but completely internally connected groups and the payoff to each member of a group is linearly proportional to the number of cooperators. They then consider the stability of a cooperative strategy (by which they mean Tit-for-Tat) as opposed to unconditional defection and show that cooperation through reciprocity becomes harder to sustain as the group size increases. They also show, in later work (Boyd and Richerson 1989), that indirect reciprocity (A helps B who helps C who helps ... who helps A) is likewise harder to sustain when the chain becomes too long. Essentially this later work is the opposite topological extreme to their 1988 paper because it represents isolated, one-dimensional *rings* of interacting players as opposed to wholly connected clusters. An important element in both these papers is that they treat the groups of players in the population as isolated and thus effectively consider only local dynamics.

At the opposite extreme, Glance and Huberman consider a problem closely related to the iterated Prisoner's Dilemma: the Diner's Dilemma (Glance and Huberman 1993, 1994). The problem statement is that a group of individuals go out to dine with the understanding that they will split the bill evenly. The dilemma is then whether to order cheaply (and so lower the overall cost) or to lash out at the expense of the others (risking that everyone else will do the same). This is a somewhat whimsical premise, but it is symptomatic of a much more general situation that arises whenever individuals are called upon to contribute their time, money, or services to "the common good." In the sense that individual cooperation comes at a cost, but group cooperation benefits all, the Diner's Dilemma is very much like a mean-field version of the N-player Prisoner's Dilemma (that is, everyone plays against the average action of the population), in which costs and benefits are anticipated over an expected time horizon. This is a powerful and elegant approach. However, in taking a purely global perspective, it misses the structural features of societies and the consequent dynamical aspect of cooperation in networks. As we shall see later, this element is an important one.

8.1.4 Evolution of Strategies

Evolution is a word that is used in many contexts, often with different meanings or connotations. For instance, the time evolution of a dynamical system is different from the evolution of cellular automata rules under the action of a genetic algorithm, which is different again from the preferential reproduction of successful strategies in a multiplayer, iterated Prisoner's Dilemma as studied, for example, by Axelrod (1980; Axelrod and Hamilton 1981; Axelrod and Dion 1988).

A distinction will be drawn here between *emergence* of a particular *behaviour* and *evolution* of a particular *strategy*. Emergence can occur in a homogeneous population, in which all players adopt the same strategy and are started in particular initial states. If, say, a small seed of initial cooperators successfully invades a sea of initial defectors, where all players utilise a Tit-for-Tat strategy, then cooperation has *emerged* in the population (see Section 8.2). By contrast, evolution can occur only in a heterogeneous population, in which different strategies exist as well as different states. Once again, the population is started with some initial distribution of states and run for one or many turns; then the resulting payoffs are used to determine which strategies performed best, and these are allowed to reproduce preferentially. This process is repeated over a number of *generations*, and if the end state is one dominated by cooperators, then cooperation is said to have *evolved* (Section 8.3).

These definitions are helpful guides but are not rigorous. For instance, evolution can occur in many ways. Lindgren (1991; Lindgren and Nordahl 1994) has adopted an approach reminiscent of the Mitchell-Crutchfield approach to cellular automata (see Chapter 7), by specifying strategies as bit strings and allowing successful strategies to reproduce, generating *new* strategies through genetic crossover and mutation. Perhaps this is true evolution, but it suffers from the difficulty of interpreting the resultant strategies in terms of any high-level language that we can then understand. A compromise is to settle for a more restricted version of evolution, in which different strategies compete for dominance in a population but themselves remain unaltered throughout the process. Whilst this may not be a realistic representation of biological evolution, it *is* a qualitatively different case to that of emergence, in that the selective force acts upon the *actions* of players (phenotypes), but it is the *distribution of strategies* (genotypes) that changes from generation to generation.

8.2 EMERGENCE OF COOPERATION IN A HOMOGENEOUS POPULATION

The first task is to understand how variations in coupling topology can affect which *behaviour* (either cooperation or defection) will dominate in an iterated Prisoner's Dilemma game in which every player utilises the *same strategy*. Specifically the following two *update rules* are considered:

1. *Generalised Tit-for-Tat.* Each player v is assigned a *hardness* h (where $0 \leq h \leq 1$) that is the same for all players, and the game is started in some initial state of cooperators and defectors. At each time step after $t = 0$, each player (v) calculates the fraction of its neighbours that cooperated on the last time step. If this fraction is greater than h, then v cooperates on its next move; else it defects. Hence, roughly speaking, $h = 0$ is equivalent to always cooperating, $h = 1$ is equivalent to always defecting, and $h = 0.5$ is equivalent to a two-player Tit-for-Tat strategy, but generalised for k players.
2. *Win-Stay, Lose-Shift.* Each player v plays against each member of its immediate neighbourhood (again, the game is started in some specified initial state) and tallies its payoff. If v scores the same as or more than the average payoff of its neighbours (including itself), then it "wins" and maintains its current action; else it "loses" and shifts to the opposite of its current action.

Both these rules are *local rules* in that each player utilises only information that is available in its *local environment*. The fraction of cooperators in the entire system is, by contrast, a *global* feature of the dynamics. As before, the local dynamics will be kept constant so that any change in the global dynamics must be due to associated changes in the coupling topology.

8.2.1 Generalised Tit-for-Tat

In any mathematical consideration of the emergence and sustenance of cooperation among competitive agents, Tit-for-Tat is the obvious place to start. Yet Tit-for-Tat, as traditionally defined, makes sense only for a two-player game. Perhaps the most obvious extension of the rule to more than two players involves each player simply playing simultaneous, but separate, two-player games against all other players (or some subset thereof). This results, however, in a lot of computationally-expensive book keeping for little apparent benefit, when each player can instead just play a single game against the average behaviour of its neighbourhood (*Generalised Tit-for-Tat*). Generalised Tit-for-Tat also sports the advantage that players can be inherently more or less cooperative, introducing some extra flexibility into the original Tit-for-Tat formulation.

However, it is a mistake to simplify things too much. Keeping in mind Huberman and Glance's result (1993), a pattern of *asynchronous updating* is adopted, in which all players are updated once each time step but in a random order, where each player can utilise the information generated by players in its neighbourhood who were updated earlier that in turn. This might seem at odds with the original Prisoner's Dilemma game, where both players decide simultaneously and find out about each other's actions only in retrospect. However, synchronous updating is quite degenerate in any kind of distributed, multiplayer context where it is virtually inconceivable that everyone would decide upon their next action at the same time, every time.

Adopting this procedure, Figures 8.1 and 8.2 summarise the results for the Generalised Tit-for-Tat game with $n = 1{,}000$, $k = 10$ and a connected seed of twenty initial cooperators.[3] For different ϕ, the steady-state fraction of cooperators (C_{steady}) exhibits different functional dependencies upon the hardness h (Fig. 8.1). In fact, thinking of h as the inverse of infectiousness (the harder the players, the less likely cooperation is to spread) and C_{steady} as the fraction of players "infected with cooperation," then the results are not dissimilar to those of the permanent-removal dynamics in Chapter 6. But there is a subtle difference. In the disease-spreading model, each individual can infect each of its neighbours with some probability ρ. So the rule is effectively If I know an infected agent, I have some chance of catching the disease. In this model, however, the rule is instead I will only do (or catch) something if so many of my friends do it first. So it's more like peer pressure than a disease, where the hardness h is a measure of how resilient people

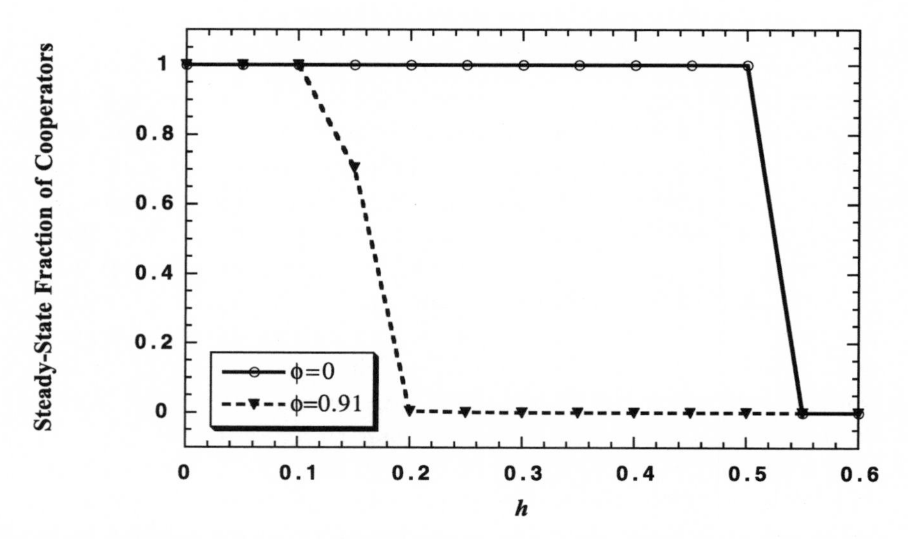

Figure 8.1 Steady-state fraction of cooperators vs. h in a Generalised Tit-for-Tat game on β-graphs, where all players have the same hardness h. Curves for the two extremal types of β-graph (1-lattice and random limit) are shown.

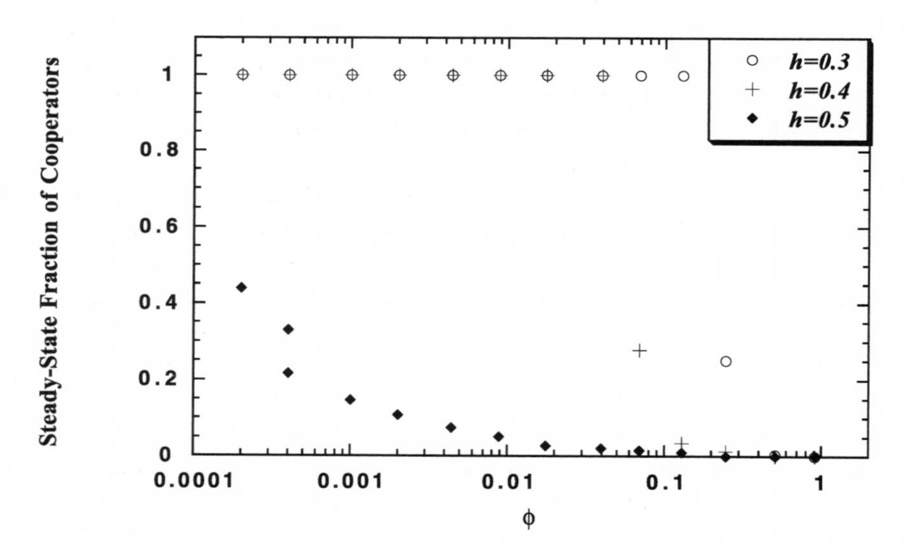

Figure 8.2 Steady-state fraction of cooperators vs. ϕ for Generalised Tit-for-Tat game on β-graphs. Curves for three values of h shown to highlight variable dependence on ϕ.

are to the pressure (that is, how *hard* they are to sway). An important consequence of this distinction is that, where disease spreading is highly sensitive to the characteristic path length of a network, cooperation is highly sensitive to the clustering coefficient. The reason is easy to see: in a clustered world, if a small group of people decide to cooperate with each other, then, within the boundaries of the cluster, each player has only other cooperators to deal with. So all the action is confined to the boundaries of the cluster, which, depending on h, either grow or contract slowly. This is the world of Axelrod's preferential mixing, in which he showed that a small seed of conditional cooperators could invade a population of unconditional defectors (Axelrod 1980). But in a random world, the initial cluster of cooperators is no longer insulated against the sea of defectors in the same way, and every player in the seed, more or less, has to fight its own battle. The result is that cooperation is rapidly crushed, unless either the initial fraction of cooperators is very high, in which case it is likely that enough cooperators can be found within even a random sample of the population, or else h is very small (for $k = 10$, the required $h \leq 0.1$), in which case cooperation is like a highly infectious disease and spreads rapidly.

For small-world graphs, the situation is more complicated. Figure 8.2 shows that for different h, no consistent dependency of C_{steady} on ϕ is apparent. For relatively "soft" populations ($h = 0.3$), small-world graphs can support the growth of cooperation (although completely random graphs still cannot), but for increasing h the value of ϕ at which cooperation collapses rapidly decreases. The overall message of these results is that the emergence of cooperation in a small-world graph requires a population that is *somewhat inclined to cooperate in the first place* (but it will then spread rapidly), whereas cooperation can thrive (but only slowly) in, say, a 1-lattice when the population is only *marginally* so inclined.

Another way to view this result is to recall Boyd and Richerson's (1988) observation that cooperation is harder to sustain in a multiplayer Prisoner's Dilemma game as n increases. Figure 8.2 appears to confirm this observation but in a novel fashion: as ϕ increases, the local neighbourhood becomes increasingly representative of the entire world, and so the effective size of population with which each player is interacting increases, *even though the average number of connections stays fixed.* Once again we encounter this phenomenon that the local and global scales collide in a small-world topology.

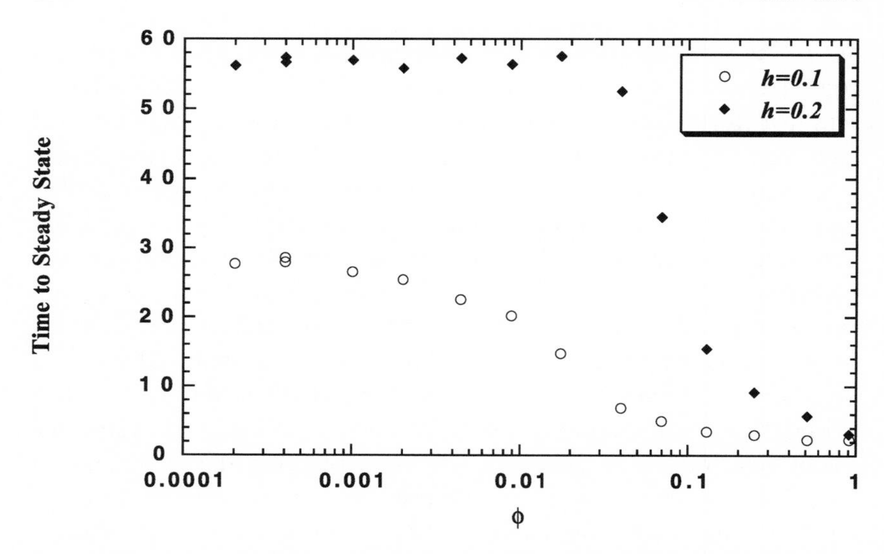

Figure 8.3 Time taken to reach a steady state vs. ϕ for Generalised Tit-for-Tat game on β-graphs. Again, different values of h yield different dependences on ϕ.

Another difference between the spread of cooperation and the spread of an infectious disease is apparent from the time taken to reach a steady state (t_{steady}). Figure 8.3 shows that t_{steady} does not satisfy the same bounds prescribed in Equation 6.5. This is probably due to a combination of two factors: the asynchronous updating allows for faster spreading than does synchronous updating (as players can react to their neighbours who updated earlier in the same round), and the crudeness of the threshold condition (h must be converted to an *integer* number of cooperating neighbours required to elicit cooperation). These factors make any bounds on t_{steady} difficult to ascertain.

As with the dynamical systems considered in earlier chapters, a natural question is whether or not the dynamics of the iterated Prisoner's Dilemma is invariant with respect to changes in the graph model. What this is really asking is whether or not the statistics L, γ, and ϕ are sufficient to capture all the intricacies of dynamical interaction that produce global statistics like $C(t)$. The agreement between the dynamics of α- and β-models up to this point has been something of a pleasant surprise and indicates that the length and clustering statistics must be at least a part of the story. Unfortunately (or fortunately, depending on how you view such things) the story *does* get more complicated. The fol-

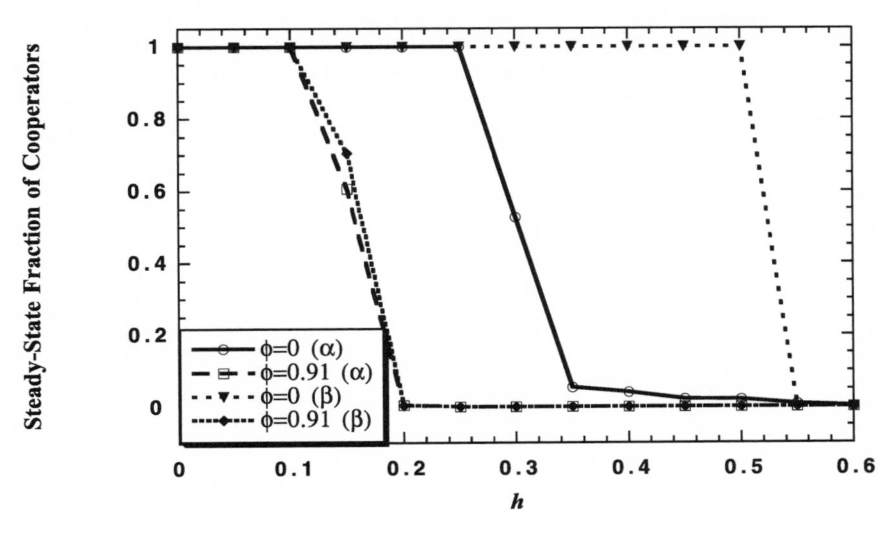

Figure 8.4 Comparison of Generalised Tit-for-Tat results for α- and β-graphs. $C(h)$ has approximately the same functional form for the random limit but is quite different at the $\phi = 0$ limit.

lowing results and those of Section 8.3 demonstrate that even the very simple dynamical systems to which the discussion has been restricted are sensitive to more than just $L(\phi)$ and $\gamma(\phi)$.

Figure 8.4 compares $C_{\text{steady}}(h)$ for the two extreme cases of zero and large ϕ for α- and β-graphs. There should be no difference between the two models at large ϕ, as this is the random limit, and it should be the case that all random graphs are more or less the same. Figure 8.4 confirms this invariance, but note that the same is not true for the $\phi = 0$ limit. The explanation again appears to be due to the crudeness of the threshold h. The α-graphs exhibit significant variance in k for $\phi = 0$ (about the same as they do for $\phi \approx 1$), whereas the β-graphs with $\phi = 0$ are necessarily k-regular. So in the α-model the same h can yield a different number of required cooperators, depending on the size of individual neighbourhoods. The same is not true for the β-model, where (at $\phi = 0$) all neighbourhoods are the same size. Again it seems that subtle differences in the local dynamics can have a major effect on the global dynamics through their interaction with the structure of the graph. More to the point, the statistics emphasised so far do not seem adequate to capture these interactions—a revelation that should not surprise anyone.

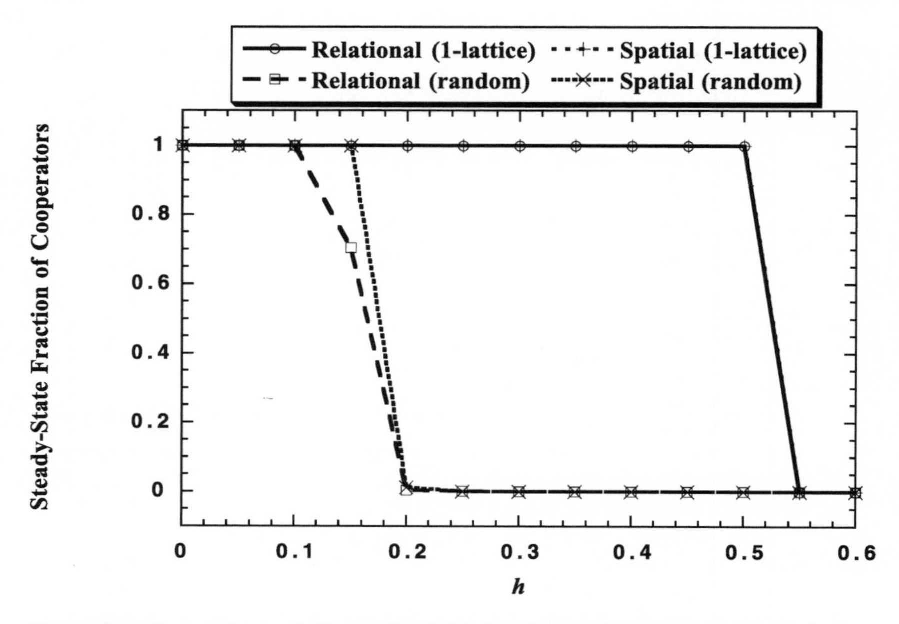

Figure 8.5 Comparison of Generalised Tit-for-Tat results for β-graphs and uniform spatial graphs. $C(h)$ has approximately the same functional form at both random and 1-lattice limits.

Before leaving the Generalised Tit-for-Tat game some spatial-graph results are in order. Here the results seem somewhat more familiar. Figure 8.5 shows that $C_{\text{steady}}(h)$ exhibits similar behaviour on spatial and β-graphs in the two extreme cases (1-lattice and random limit). Figure 8.6 strengthens this impression, although it is hard to compare quantitatively with Figure 8.2 (as ξ and β cannot be expressed in terms of a single parameter). However, Figure 8.7 indicates that (for at least one value of h) t_{steady} is determined in a straightforward manner by $L(\xi)$, in contrast to the β-graph results above.

8.2.2 Win-Stay, Lose-Shift

Despite the discrepancies noted above, the results of the Generalised Tit-for-Tat model are not all that different from those in Chapter 6. In fact, it seems reasonable to think of cooperation as spreading through a society from some small, initial seed in much the same fashion as a disease, where the effectiveness of the spreading is governed by how predisposed the citizens are to cooperation in the first place. Win-Stay,

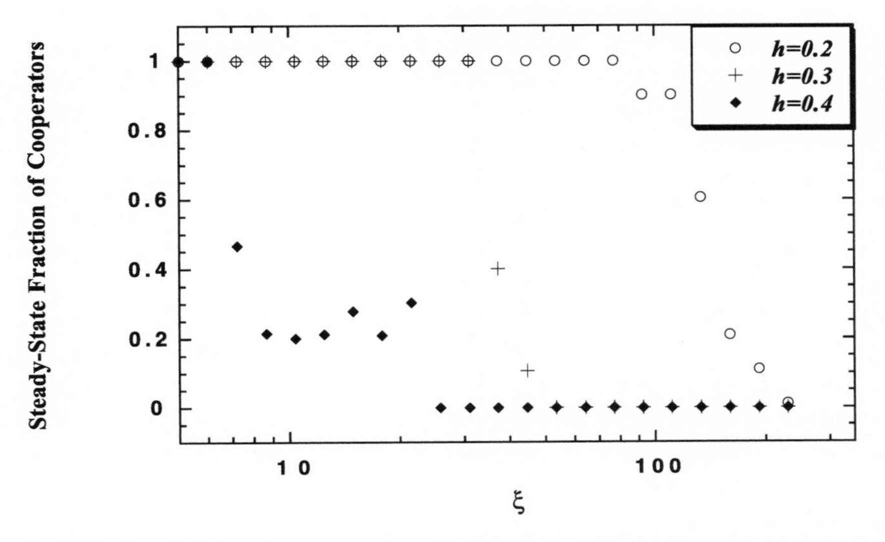

Figure 8.6 Steady-state fraction of cooperators vs. ξ for Generalised Tit-for-Tat game on uniform spatial graphs. Curves for three values of h are shown to highlight variable dependence on ξ.

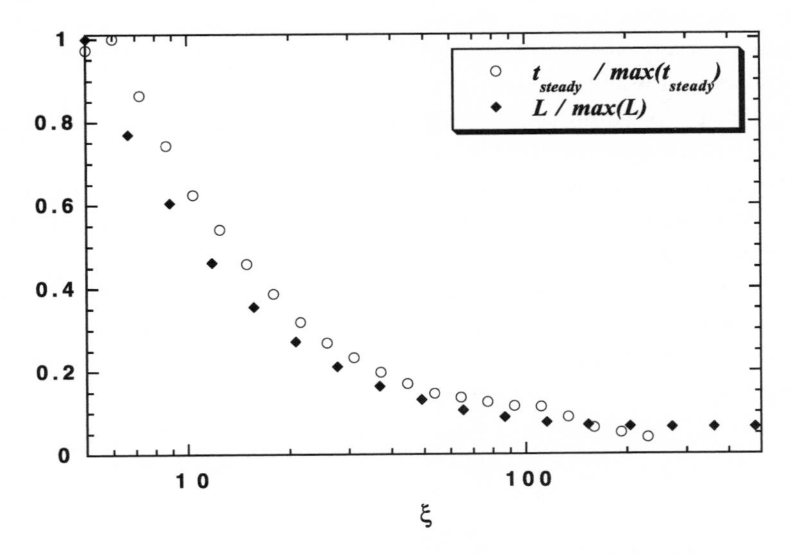

Figure 8.7 Time taken to reach a steady state vs. ξ for Generalised Tit-for-Tat game on uniform spatial graphs (with $h = 0.2$) shows similar functional form to $L(\xi)$. (Both t_{steady} and L have been scaled by their maximal value.)

Lose-Shift represents a different kind of psychology altogether. Players operating according to this strategy are not predisposed to any action in particular: only that which yields reward and avoids punishment. Hence Win-Stay, Lose-Shift, also known as Pavlov (after its salivatory predecessor), produces dynamics that are not reducible to mere spreading. In fact, some rather more complicated things can happen. For instance, if a single cooperator v is introduced into a sea of initial defectors (where everyone employs Win-Stay, Lose-Shift), then the cooperator will be soundly trounced by all its neighbours and will switch to defection in the next round. Meanwhile, all its neighbours will happily continue to defect as this action served them well in suckering the lone cooperator. However, some of *their* neighbours (that is, the elements of $\Gamma^2(v)$) did not get to sucker v and so scored less than the players who did. Consequently they do less well than the average payoff and decide to change their action to cooperation, not realising that it was cooperation that got v into trouble in the first place.

The resulting dynamics is sufficiently complicated that it never (so far as any simulations could detect) settles down into either a steady state or periodic orbit.[4] However, the *statistics* of the system (that is, the fraction of cooperators $C(t)$) do settle down to asymptotically stable values. Curiously enough, the system seems to settle down to the same value of C for all n, all initial states, and even for most ϕ. Figure 8.8 shows that, for large ϕ, cooperation becomes slightly *more* prevalent (but—look at the vertical scale—not by much), in contrast to the results of the previous section, where cooperation was harder to sustain at large ϕ. Topology still plays a role, however. Figure 8.9 shows that the time taken for the asymptotic state to be reached varies dramatically with ϕ, in much the same way as we have seen before. Furthermore, t_{asymp} grows linearly with n for $\phi = 0$ (Fig. 8.10) and logarithmically for $\phi \gtrsim 0.002$ (Fig. 8.11). It is no coincidence that $\phi \approx 0.002$ is the smallest ϕ at which logarithmic length scaling can be observed: $0.002 = 1/500$ and $n = 500$ is the smallest graph of the range over which the scaling results were compiled, so $\phi = 0.002$ is the value at which the smallest graph can be expected to contain one shortcut. Hence one would expect that, in the limit $n \to \infty$, any $\phi > 0$ would result in logarithmic scaling of t_{steady}—an identical condition to that for length scaling of relational graphs (see Chapter 4). Once again, the length characteristics of the underlying graph appear to be important in determining the timescale on which (at least some) global dynamics occur.

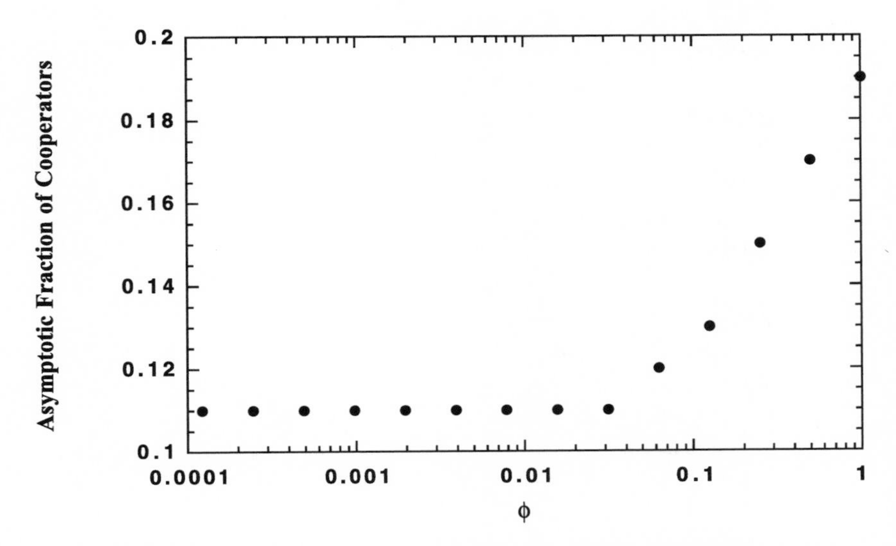

Figure 8.8 Asymptotic fraction of cooperators vs. ϕ for the Win-Stay, Lose-Shift strategy on β-graphs ($n = 1{,}000$, $k = 10$).

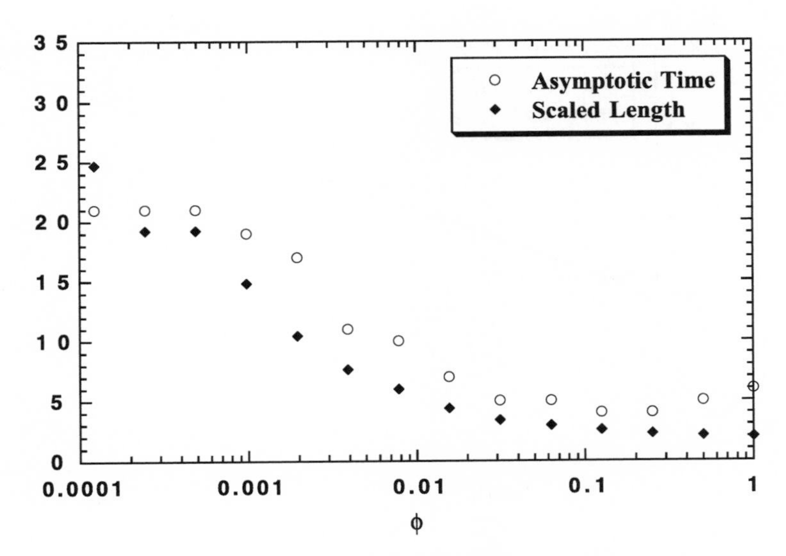

Figure 8.9 Time taken to reach asymptotic steady-state vs. ϕ for Win-Stay, Lose-Shift strategy on β-graphs ($n = 1{,}000$, $k = 10$), compared with a scaled version of $L(\phi)$.

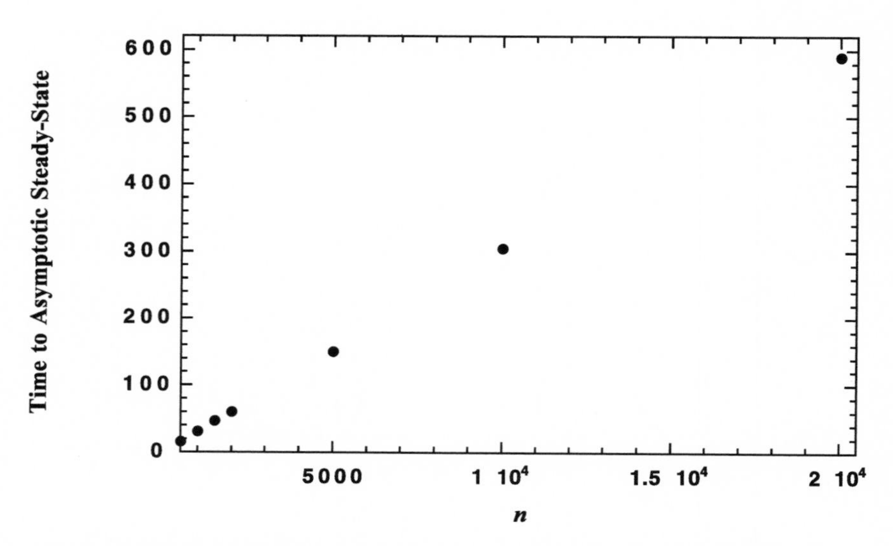

Figure 8.10 Scaling of t_{asymp} with respect to n is clearly linear for Win-Stay, Lose-Shift strategy on β-graphs with $\phi = 0$ ($k = 10$).

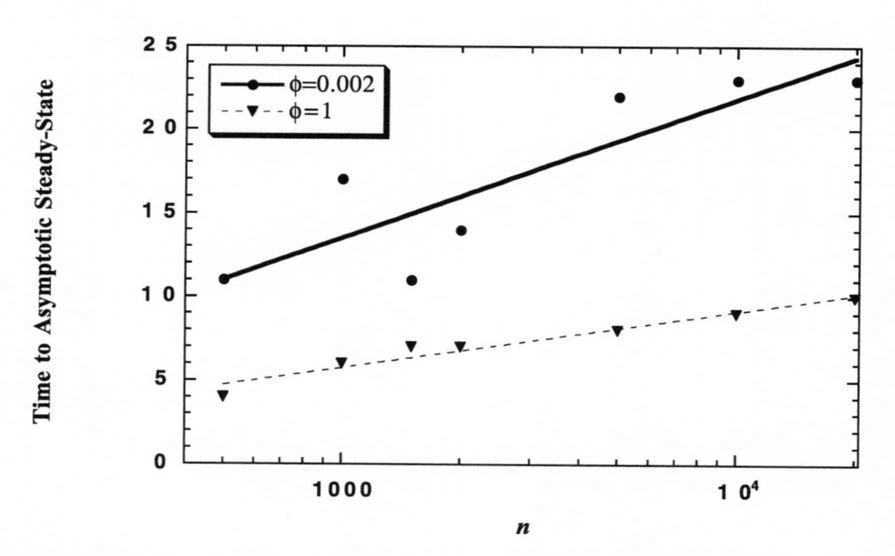

Figure 8.11 Scaling of t_{asymp} with respect to n is logarithmic for Win-Stay, Lose-Shift strategy on β-graphs with $\phi = 0.002$ and $\phi = 1$ ($k = 10$). Logarithmic curve fits are shown for clarity.

8.3 EVOLUTION OF COOPERATION IN A HETEROGENEOUS POPULATION

Evolution is a natural next step to take from the emergence of cooperation: that is, how is the evolution of successful strategies determined by the structure of the population, in which multiple strategies can now coexist? Instead of playing just a single, iterated, multiplayer Prisoner's Dilemma game, one can play a series of such games (generations), between each of which the strategy that each player uses may change, depending on some selection criterion. This is a more complicated dynamical system than earlier examples, because it involves two kinds of dynamics: what we might call *behavioural dynamics*, which describes what happens to the states of the players during any given generation, and *strategy dynamics*, which describes how the update rules change over the course of generations.

As mentioned earlier, the strategy dynamics considered here is quite restrictive in that the strategies do not themselves change. Rather, an initial set of strategies is assigned in some fashion, after which individual players can change from their current strategy to another available option, based on a selection criterion (which can be thought of as a *meta-strategy*), so what evolves effectively is the *mix* of strategies in the population. Specifically, the initial range of strategies consists of Generalised Tit-for-Tat rules with different h, and the meta-strategy is called *Copycat*: at the end of each generation, every player assesses their tallied score (over all rounds of the entire game) and that of each of their neighbours; each player then adopts the *strategy* of the highest scoring player in their neighbourhood (which could be themselves) for use in the next game. Hence successful strategies reproduce preferentially, and unsuccessful strategies are eliminated site-by-site and may eventually disappear from the population altogether.

Two scenarios are considered, each of which is then allowed to evolve for 100 generations: (1) A small fraction (0.1 of the population) of *unconditional cooperators* ($h = 0$), embedded in a population of *unconditional defectors* ($h = 1$). (2) A small fraction (again 0.1 of the population) of *conditional cooperators* ($h = 0.5$), embedded in a population of unconditional defectors.

Figures 8.12 and 8.13 indicate that, once again, the global dynamics do not behave consistently for the α- and β-models. Again the relatively high variance of the degree in the α-graphs appears to have a debilitating

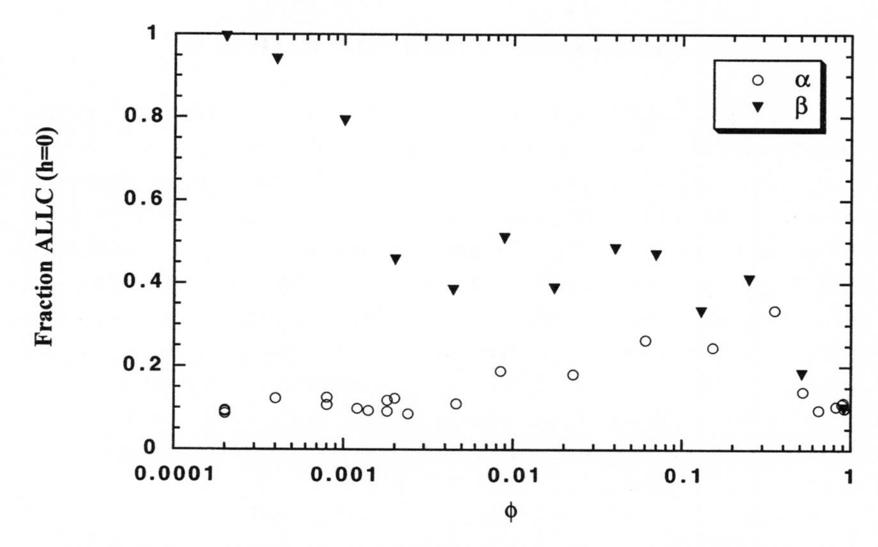

Figure 8.12 Evolved fraction of unconditional cooperators ($h = 0$) from an initial seed in a population of unconditional defectors ($h = 1$). A comparison between α- and β-graphs is shown vs. ϕ.

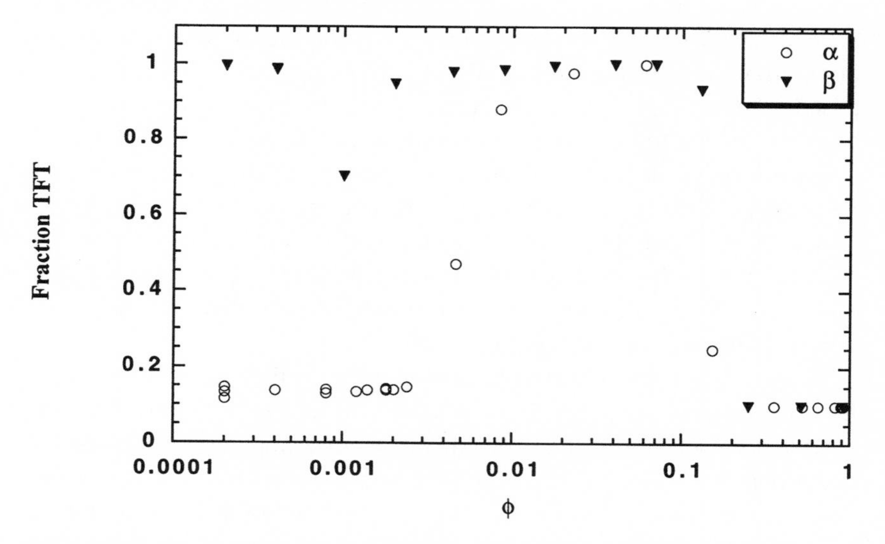

Figure 8.13 Evolved fraction of conditional cooperators ($h = 0.5$) from an initial seed in a population of unconditional defectors ($h = 1$). A comparison between α- and β-graphs is shown vs. ϕ.

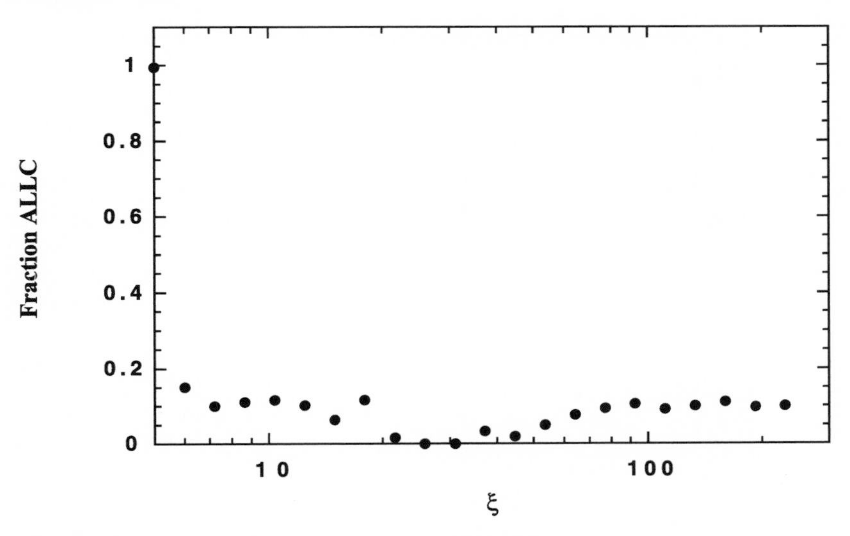

Figure 8.14 Fraction of unconditional cooperators ($h = 0$) vs. ξ in a sea of unconditional defectors ($h = 1$) on uniform spatial-graphs. Initial seed size of unconditional cooperators = 0.1 of population.

effect on the spread of cooperation for low ϕ. An interesting deviation from previous results, however, is that cooperation does not necessarily perform worse for higher ϕ, at least in the case of the α-graphs (note the hump in Fig. 8.13), where both the unconditional and conditional cooperators fare better in at least some small-world graphs than in either caveman-like or random graphs.

In contrast, the uniform spatial model shows that cooperation appears to die rapidly and with little respite as ξ is increased. Figures 8.14 and 8.15 show that unconditional defection dominates both alternative strategies for any $\xi > k$. It would be interesting to see if this were also the case in a more heterogeneous environment, where a greater range of strategies were able to compete.

8.4 MAIN POINTS IN REVIEW

In summary, the introduction of the topologies of Part I to games of cooperation seems to have a significant impact on both the emergence of cooperative behaviour in a homogeneous population and the evolution (or, more accurately, preferential reproduction) of cooperative strategies in a heterogeneous population. Unfortunately it is difficult to draw

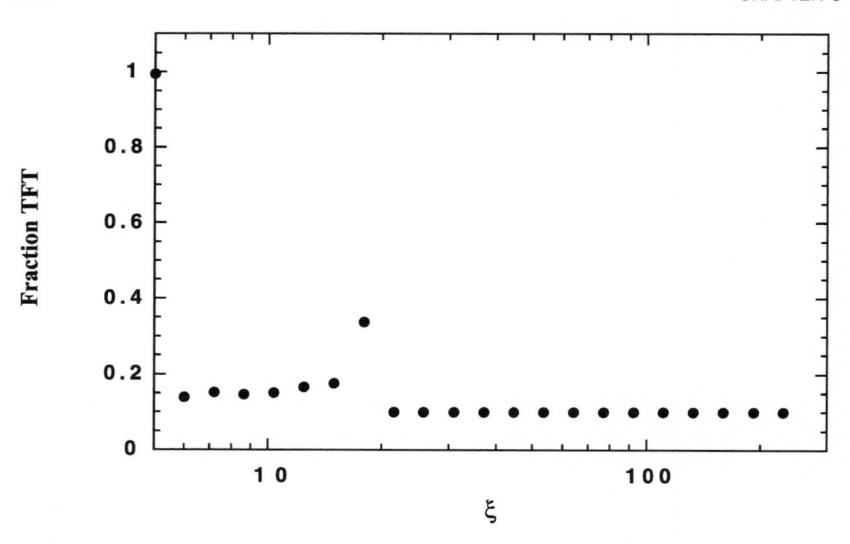

Figure 8.15 Evolved fraction of conditional cooperators ($h = 0.5$) vs. ξ in a sea of unconditional devectors ($h = 1$) on uniform spatial graphs. Initial seed size of conditional cooperators $= 0.1$ of population.

much in the way of generally applicable lessons from this, except that in the case of Generalised Tit-for-Tat strategies, cooperation *tends* to do worse in poorly clustered graphs such as random graphs. This seems to be because cooperation, as defined by the Prisoner's Dilemma and strategies like Tit-for-Tat, relies for its success upon a *group* of cooperators banding together against the evils of an uncooperative world and scoring points by cooperating with *each other*. Once a few defectors can infiltrate this seed, by way of shortcuts, then the fledgling cooperation rots from the core out and collapses. In the context of the evolution of strategies, however, this does not always seem to be the case, as (recall Fig. 8.13) cooperation seems sometimes to evolve preferentially in small-world (but not random) graphs.

There is, however, a more positive spin to this story: for at least some range of h, cooperation does just fine in a small world (although not in a random one). This may have implications for organisational design, where both the efficient transmission of information *and* generally cooperative behaviour are advantageous to an organisation's performance. Solving optimisation problems like this by varying the connectivity of networks is not an approach that has received much attention, but it may turn out to be useful in a whole range of applications.

9

Global Synchrony in Populations of Coupled Phase Oscillators

One of the distinguishing features of all the dynamical systems considered up to this point is that they are *discrete*. That is, each element of the system occupies one of only a finite number of states: on or off; cooperating or defecting; susceptible, infected, or removed. A qualitatively different kind of dynamical system is one that can occupy a *continuum* of states. Often these two classes of system—discrete and continuous—exhibit analogous but distinct dynamical properties, so it would be interesting to see if the behaviour of a simple continuous system is as responsive to network structure as its discrete counterparts. Enter coupled oscillators, or at least a very specific, simple class of coupled oscillators, about which much is known in more traditional coupling topologies such as mean-field coupling and coupling on a ring. Once again the goal is not to produce any *general* theory linking coupling topology to dynamics, but rather to choose specific, simple systems so that the changes in global dynamics due to changes in coupling topology can be isolated and thus better understood. Even so, as we have already seen, things get quite complicated enough.

9.1 BACKGROUND

The subject of coupled oscillators is a large and impressive one, infiltrating almost as many fields as graph theory and with an even longer history. The first recorded observation of two oscillators operating in an interdependent (coupled) fashion is in a letter from Christian Huygens to his father in 1665 (Huygens 1893). Huygens, who was sick in bed at the time, noticed that two of his pendulum clocks, which were hanging next to each other on the wall, kept identical time and returned to a synchronous, anti-phase-locked state even if the pendulum of one of them was disturbed. Conversely, when he placed them on opposite sides of the room, they gradually drifted apart. He concluded that clocks sufficiently close to each other were *coupled* through small vibrations in the wall.

Since the seventeenth century, theoretical and experimental work relating to coupled oscillators has impinged upon fields as diverse as neurophysiology (Kopell 1988; Schuster and Wagner 1990a, 1990b), population biology (Buck 1988), physics (Hadley et al. 1988), and even women's menstrual cycles (Russell et al. 1980). Most of this development has occurred in the past thirty years, starting from the highly original work of Arthur Winfree (1967), who, soon after his undergraduate studies at Cornell, investigated the properties of populations of *weakly coupled* oscillators that were both *self-sustaining* and *stable* with respect to external perturbations. These characteristics, intended to represent driven oscillators with nonlinear damping that are often observed in biological systems, enable a substantial reduction in the number of variables that need to be considered. That is, the *full* state of the oscillator can be thought of as a point in some high-dimensional phase space, but its motion is restricted to within small perturbations of a single *limit cycle* of fixed amplitude. Hence the oscillator's state can be represented by a single variable: its *phase* on the limit cycle (θ). Moreover, the weak-coupling assumption ensures that the influence of other oscillators, due to the coupling, affects only the phase of the oscillator and not its amplitude. These restrictions allowed Winfree to describe such *phase oscillators* using the simple equation

$$\theta_i = \omega_i + Z(\theta_i) \sum_{j=1}^{n} X(\theta_j), \tag{9.1}$$

where

$\theta_i =$ the phase of the i-th oscillator,
$\omega_i =$ the intrinsic frequency of the i-th oscillator,
$Z =$ the sensitivity function,
$X =$ the influence function.

Winfree then showed that when a certain relationship was satisfied between the width of the frequency distribution $g(\omega)$ and the strength of the coupling interaction (determined by the sensitivity and influence functions), a dramatic transition to a globally entrained state would occur. In this state, all oscillators would act as one, possibly at a frequency that none of them had possessed individually. Furthermore, the entrained-state distribution of *phases* would be related to the original distribution of intrinsic frequencies. This is analogous to a pack of runners running around a track. They all pace off each other, so even though some are inherently faster than others, they all run together. Nevertheless the better runners tend to lead the pack, and the less able just hang on to the end. If the motivation to stay with the pack fades too much or the range of natural abilities is too broad, the pack necessarily breaks up and entrainment is lost.

This idea of Winfree's—that much could be understood about large populations of biological oscillators with simple and attractive equations—started an avalanche of work (see Strogatz 1994 for a user-friendly review), perhaps the most significant, single contribution to which was made by the statistical physicist Yoshiki Kuramoto (1975), who considered a simplified version of Winfree's equation:

$$\dot{\theta}_i = \omega_i + \frac{\lambda}{n} \sum_{j=1}^{n} \sin(\theta_j - \theta_i), \tag{9.2}$$

where λ (the coupling strength) is constant for all oscillators. The Kuramoto equation is a very special case of Winfree's model, where the coupling term is both *symmetric* and depends only on phase *difference*, not phase itself. Employing these features to great effect, Kuramoto managed to derive a self-consistency condition for the system, in terms of a statistical *order parameter*, out of which popped both the correct value of the *critical coupling strength* (λ_c) (at which macroscopic clusters of frequency-locked oscillators appear) and the functional form of the order parameter of the system for $\lambda > \lambda_c$. Specifically, if the population of oscillators is visualised as n points moving around a circle (like the runners on the track), the position of the *centroid* of the population can be expressed as a vector (see Fig. 9.1):

$$Re^{i\Psi} = \frac{1}{n} \sum_{j=1}^{n} e^{i\theta_j} \tag{9.3}$$

Kuramoto's *phase-order parameter* R is just the magnitude of this vector, and Ψ is its angular coordinate. Hence a uniform, random distribution of the oscillators around the ring corresponds to $R = 0$, and $R = 1$ corresponds to perfect, universal synchrony. Kuramoto showed that

$$\lambda_c = \frac{2}{\pi g(0)},$$

where $g(\omega)$ is the distribution of intrinsic frequencies, which he assumed to be symmetric and unimodal (single-humped). For $\lambda < \lambda_c$ no synchrony is possible,[1] and $R(t)$ decays rapidly to a small, residual value

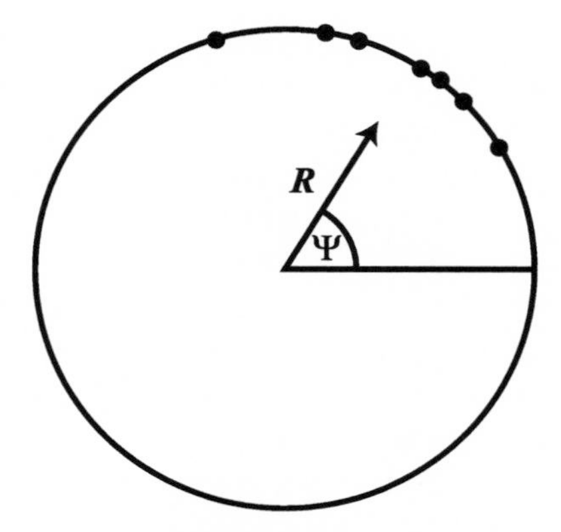

Figure 9.1 Schematic illustration of Kuramoto's phase-order parameter R.

of $O(1/\sqrt{n})$. But for $\lambda > \lambda_c$, synchrony emerges spontaneously and $R(t) \to R_\infty$, about which it continues to undergo $O(1/\sqrt{n})$ fluctuations indefinitely. Kuramoto also showed that for $\lambda \gtrsim \lambda_c$, $R_\infty \propto \sqrt{(\lambda - \lambda_c)}$, a prediction that was later confirmed (along with λ_c itself) by numerical results.

Both Winfree's and Kuramoto's work was done using what is known as a *mean-field* theory, where each oscillator is coupled directly and equally to all others and so is coupled effectively to the *mean* of all the individual influences. Mean-field theory is certainly a natural starting point for large populations because not only is it the most analytically tractable topology around, but it is a reasonable model of some systems. Furthermore, utilising a mean-field approach does not by any means render the problem simple, and Kuramoto's analysis raised a number of subtle questions that theorists spent the next twenty years answering (see Strogatz and Mirollo 1991 for some progress on this front).

Nevertheless, if the phenomenon of oscillator entrainment is to be understood in naturally occurring populations, then analogues of the mean-field results must be sought for topologies that are *sparsely coupled*, meaning that each oscillator is connected only to a *small fraction* of the total population. The predominant approach to this challenge has been to consider phase oscillators coupled on low-dimensional lattices. Motivated by the neural system of the lamprey, Kopell and Ermentrout (1986) considered one-dimensional chains of oscillators coupled via phase differences to their nearest neighbours and explained the existence of frequency locking and travelling waves. They also analysed the effects of "multiple coupling" (that is, increasing k beyond $k = 2$; Kopell et al.

1990) and occasional "long-range" connections (Ermentrout and Kopell 1993), which helped to explain how some travelling waves could have wavelengths on a global length scale.

Work has also proceeded on oscillator lattices of two and three dimensions. Sakaguchi et al. (1987) showed that lattice dimension has a dramatic effect on the ability of a nearest-neighbour coupled system to synchronise: the lower the dimension, the greater the associated difficulty. Daido (1988) later used a renormalisation argument to determine a lower bound on the lattice dimension required for macroscopic clusters of oscillators to become mutually entrained, that is, oscillate with the same frequency. Specifically he showed that, for any distribution with a finite variance (of which the normal distribution is the generic example), global entrainment is only possible for $d \geq 2$—an observation made earlier by Sakaguchi et al. All these results were explained and amplified by Strogatz and Mirollo (1988), who showed (under quite weak conditions on $g(\omega)$) that global-sized blocks of entrained oscillators are impossible in any dimension, in the infinite limit of n, but that global entrainment might still be possible through macroscopic spongelike structures in dimensions greater than one.

Finally, a small amount of research has been conducted on what might be called "nonstandard" topologies. Satoh (1989) performed some numerical experiments comparing the self-entraining capabilities of van der Pol oscillators (which are not phase oscillators, but which are similar in some important respects) on two-dimensional lattices and sparse, random graphs. Satoh found that the system became globally entrained much more effectively in the random case. In fact, Matthews et al. (1991) note that the coupling strength required for global frequency locking in a random net is virtually the same as that required in the mean-field case. An even more unusual topology was considered by Lumer and Huberman (1991), who conducted an analysis similar to Daido's but on what is effectively a Cayley tree with variable branching, in which the coupling strength drops exponentially with distance on the graph. Instead of determining a critical *dimension* based upon the frequency distribution, they derived a critical relationship between the distribution, the branching of the tree, and the rate of exponential decline of the coupling force with distance.

However, the paper most closely related to the problem at hand is a comparison by Niebur et al. (1991), of three different connectivity schemes on a two-dimensional lattice. The first is the traditional nearest-neighbour lattice topology, and the second is an all-to-all coupling, where the coupling strength decreases with distance as a Gaussian. But the third scheme is the most intriguing: a "sparse coupling" arrangement in which the *probability* of connection declines as a Gaussian, but in which every connection that is actually made is of equal strength, regardless of its length. Niebur et al. noted that correlations in oscillator phase decreased with the distance between oscillators in the first two coupling schemes but not in the sparse coupling scheme, indicating that it might be qualitatively different—a result on which more light will be shed at the end of this chapter.

9.2 KURAMOTO OSCILLATORS ON GRAPHS

Kuramoto oscillators are not completely understood in any topology, but they have at least been well explored in a mean-field context, in low-dimensional lattices, and in some other isolated cases. This state of affairs is actually a better one than for any of the dynamical systems considered in previous chapters. But still no work to date has focused on the coupling topology as an *explicit mechanism for changing the dynamics*—that is, as another parameter that can be adjusted along with all the others in a more or less continuous fashion. Once again, this is easy to do by recasting Kuramoto's problem on β-graphs and examining the interval between the 1-lattice and random extremes. Specifically, consider a system of n Kuramoto oscillators, for which the coupling term in Equation 9.2 is summed only over the k_i immediate neighbours of the oscillator i, as specified by some underlying graph. Kuramoto's order parameter R can then be measured as a function of the coupling strength λ and the relevant graph parameter (ϕ for relational graphs and ξ for spatial graphs).

Before doing this, however, a benchmark is required against which to compare the results. The obvious standard is that set by the mean-field case, whose statistical properties can be corroborated with the results cited above. Figure 9.2 shows $R(\lambda)$ for a system of one thousand oscillators for which ω_i and the initial θ_i are taken to be uniformly distributed on $(-1/2, 1/2)$ and $(-\pi/2, \pi/2)$, respectively. As expected, for small λ, $R = O(1/\sqrt{n})$, and, in the large-n limit, $\lambda_c \approx 0.64 = 2/\pi$, both of which agree with Kuramoto's infinite-n predictions (where $g(0) = 1$ in this case). Also, in the particular case of a uniform distribution $g(\omega)$, a *mutually entrained cluster* the size of the entire system appears suddenly at $\lambda = \lambda_c$. That is, for $\lambda < \lambda_c$, the frequencies of the oscillators are

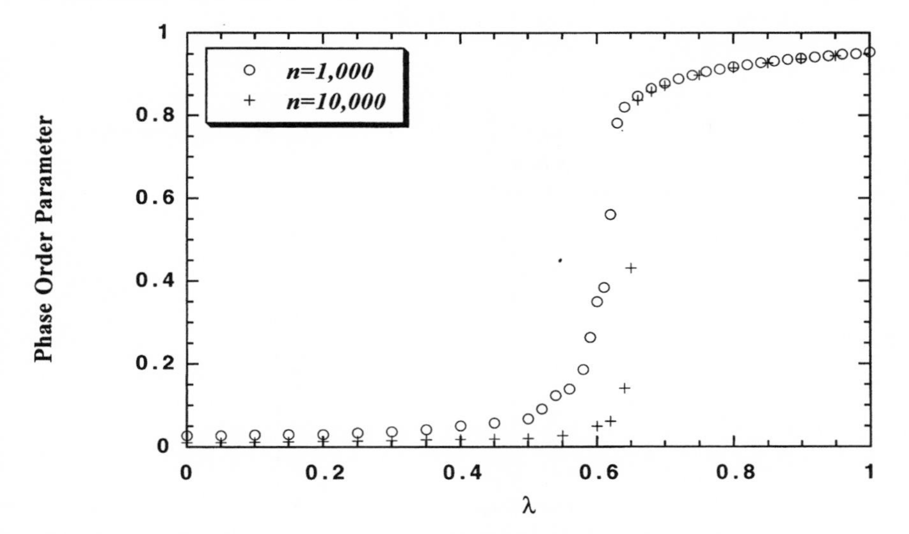

Figure 9.2 Phase-order parameter R vs. coupling strength λ on the complete graph. At large n, the mean-field, infinite n approximation is approached.

randomly distributed and no macroscopic clusters exist, but for $\lambda \geq \lambda_c$, almost all oscillators exhibit the same characteristic frequency (although they may still be distributed over a range of phases). In practice, it is necessary to set some tolerance that then determines whether or not two oscillators are counted as being in the same cluster. A useful approach is that of Sakaguchi et al. (1987), who define a *frequency-order parameter* $E = n_E/n$, where n_E is the number of oscillators in the largest cluster. Two oscillators are counted as being in the same cluster if their average frequencies differ by less than $\Delta\omega = 1/T$, where T is the time interval over which the integration is run. In fact, it does not make much difference what value is chosen for $\Delta\omega$, as long as it is small compared with the width of the frequency distribution (in all the results that follow, $\Delta\omega = 0.001$). Figure 9.3 indicates that, in the mean-field case, as n becomes sufficiently large, a discontinuous jump to global entrainment does indeed occur at λ_c for the complete graph. As Figures 9.2 and 9.3 show basically the same phenomenon, and because global entrainment is a slightly more general phenomenon than synchrony (entrainment can occur in the absence of synchrony but not the reverse), all the following results are presented in terms of the frequency-order parameter E.

First, the performance of a β-graph at its random limit ($\beta = 1$) can be compared with that of the mean-field case. Figure 9.4 shows that, for $n =$

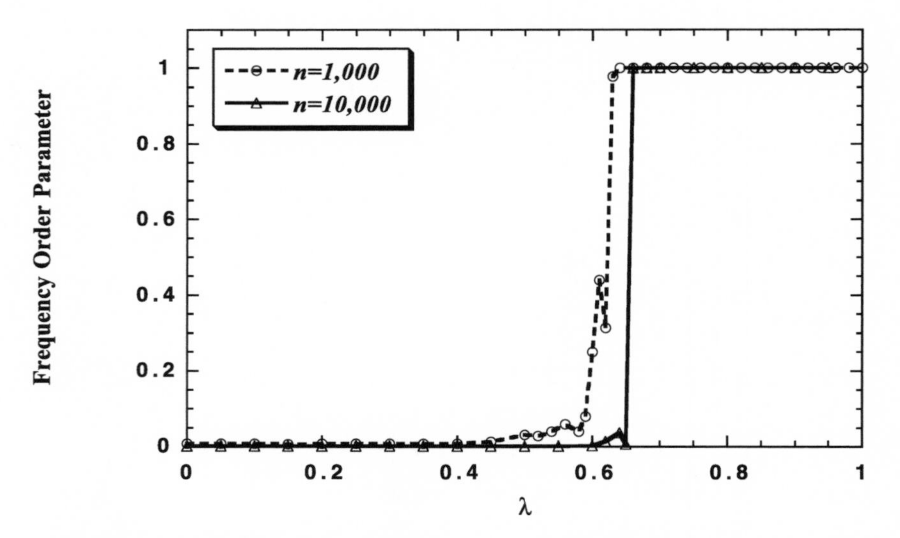

Figure 9.3 Frequency-order parameter E vs. coupling strength λ on the complete graph.

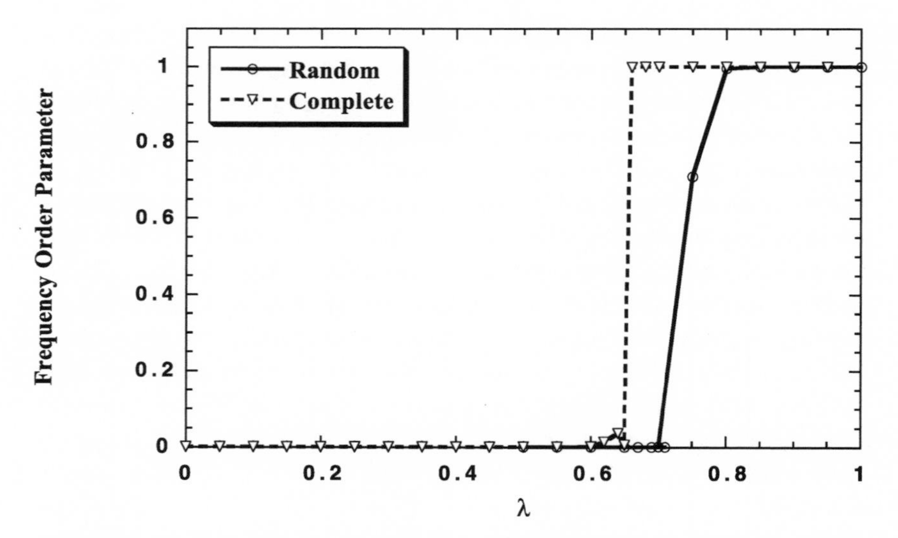

Figure 9.4 Comparison of E vs. λ between the complete graph ($n = 10{,}000$) and a β-graph at its random limit ($n = 10{,}000$, and $k = 10$).

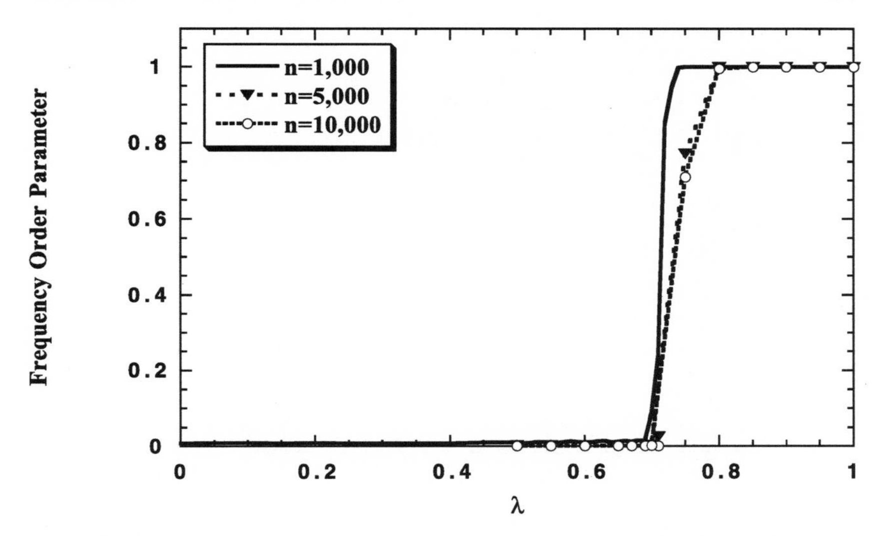

Figure 9.5 Frequency-order parameter E vs. λ for a β-graph at its random limit ($\beta = 1$) for ($n = 1{,}000$, $n = 5{,}000$, and $n = 10{,}000$). As n increases, $E(\lambda)$ appears to approach an invariant curve.

10,000, the randomly connected system exhibits a similar phase transition to that of the mean-field system despite the fact that it has a thousand times fewer edges (as $(n \cdot k)/n(n-1) \approx 0.001$). Figure 9.5 indicates that the random-graph transition also appears to be independent of n, in the sense that it appears to approach an invariant curve in the large-n limit. The transition, however, occurs at a value of λ_c that is distinctly greater than λ_c for the complete graph.

This apparent invariance of the phase transition with respect to n might be understandable in terms of a random sampling argument similar to that used in Chapter 2. There, a theorem due to Huber (1996) showed that the median shortest path length can be estimated by random sampling where the sample size depends only on the required precision of the estimate, not on n. In a uniform distribution in the infinite-n limit, the mean and median frequency will be identical, and so any oscillator could lock to the average frequency through its coupling to only a random sample, whose minimum required size would not depend on n. As the sample size is increased, so is the precision of the sampling estimate, and hence the frequency locking. If true, this would imply that, as k is increased (but still keeping $k \ll n$), then λ_c for the random graph will approach that of the complete graph. This assertion is supported by Figure 9.6, which shows that for $n = 10{,}000$ and $k = 20$ the random-graph

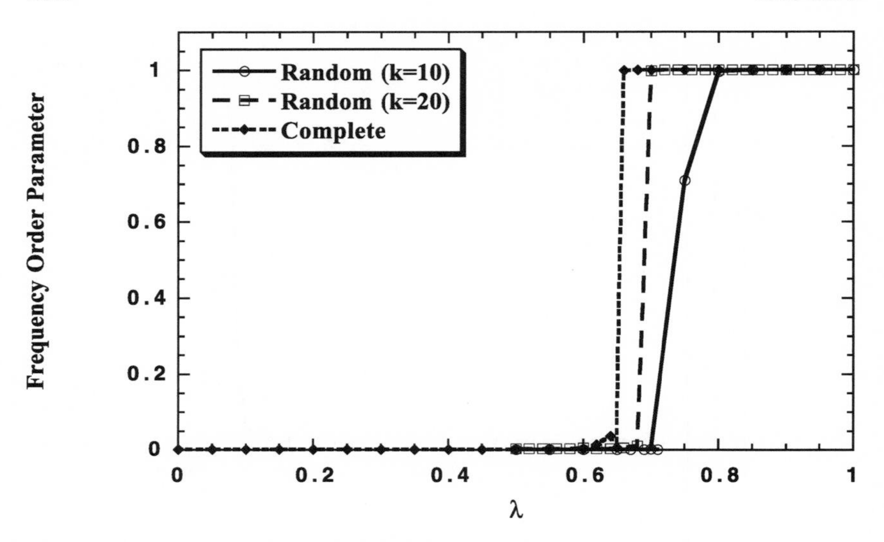

Figure 9.6 Comparison of frequency-order parameter E vs. λ between the complete graph ($n = 10{,}000$) and β-graphs at their random limit for $k = 10$ and $k = 20$. Increasing k results in improved performance of the random graphs.

phase transition occurs for significantly lower λ_c than the corresponding transition for $k = 10$. It seems plausible at least that, for some larger k (which is still small compared with n), the random graph will perform comparably to the complete graph, but with far fewer edges required.

If random graphs are capable of similar performance to that of the complete graph, the natural next question is whether or not they have to be completely random; that is, can graphs with $\beta < 1$ exhibit the same phase-transition properties that occur for $\beta = 1$? Figures 9.7 and 9.8 indicate that the answer is both yes and no: "yes," because rapid transitions to synchrony do occur in β-graphs that possess even a small fraction of shortcuts (for example, the $\phi = 0.12$ curve in Fig. 9.7); "no," because these transitions occur less rapidly and at greater λ_c for smaller ϕ.

Figure 9.9 provides a different perspective on the issue, showing that rapid transitions to global entrainment can be generated for *fixed* λ by increasing ϕ. Recall that, up to this point, all the transitions in Kuramoto oscillators resulted from changing the coupling strength λ. Hence we can think of ϕ as another parameter in the dynamics, equivalent to λ in the sense that either can be responsible for a rapid transition to global order. However, just as $E(\lambda)$ exhibits different functional forms depending on ϕ (Fig. 9.7 and 9.8), so does $E(\phi)$ depend on λ. The result is that, for any ordered pair (λ, ϕ) in the parameter space of the system, global

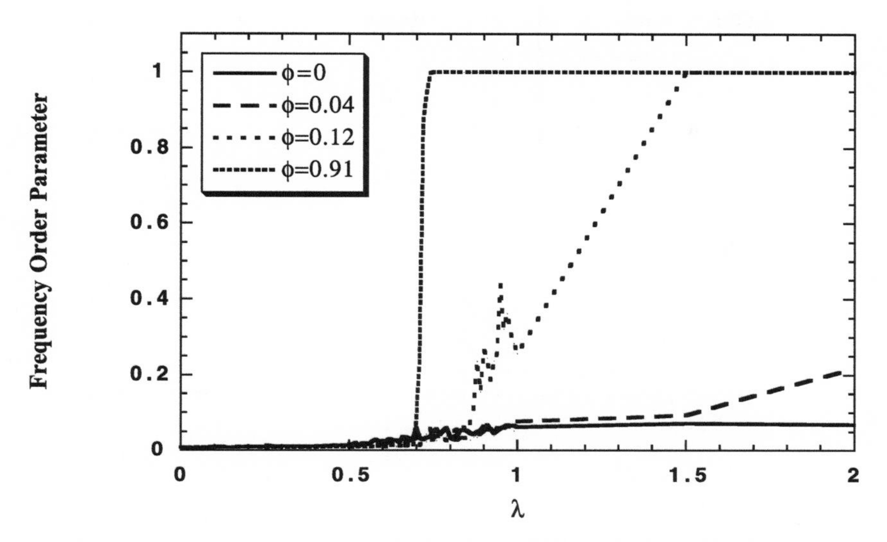

Figure 9.7 Frequency-order parameter E vs. coupling strength λ for Kuramoto oscillators on β-graphs with various values of ϕ ($n = 1{,}000$, $k = 10$).

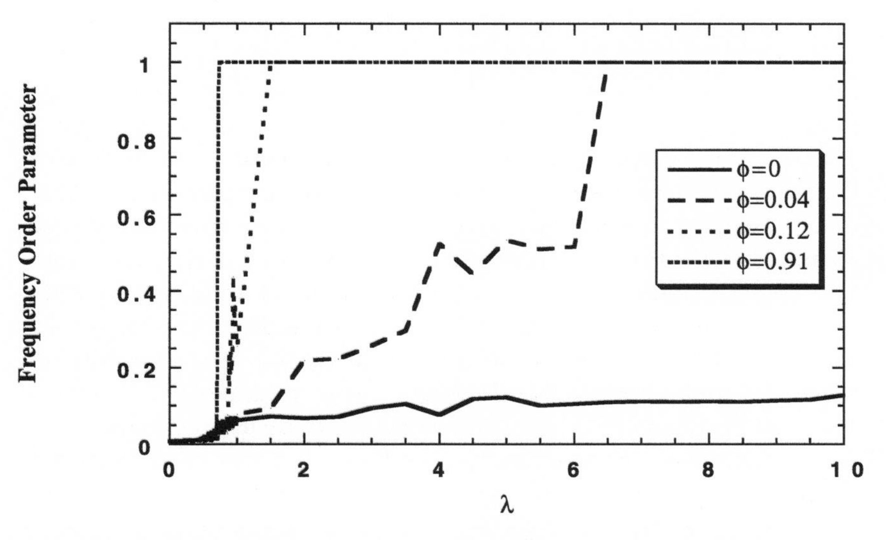

Figure 9.8 Frequency-order parameter E vs. λ for various ϕ over a larger range of λ. Transitions to global entrainment occur at higher λ_c for smaller ϕ.

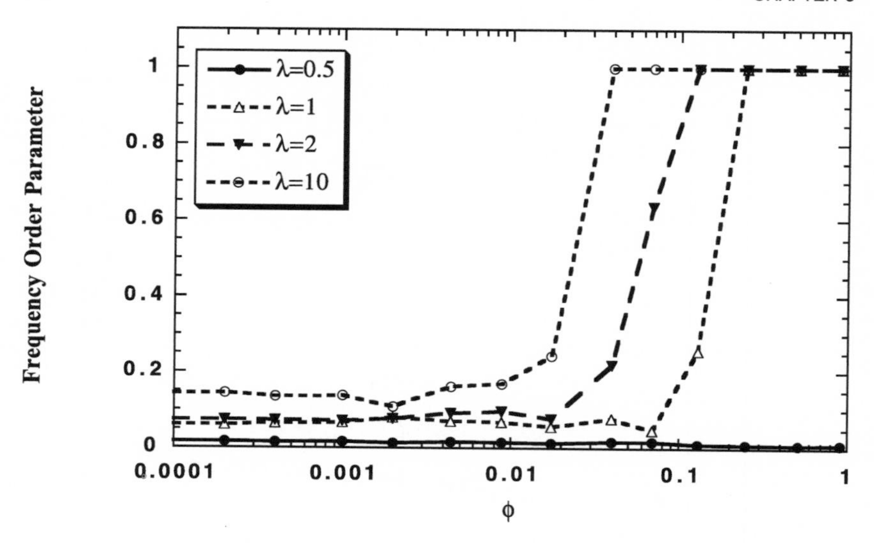

Figure 9.9 Frequency-order parameter E vs. ϕ for Kuramoto oscillators on β-graphs at various values of λ ($n = 1{,}000$, $k = 10$).

entrainment may be generated more easily by varying one as opposed to the other. This really should come as no surprise by now: we have already seen several instances in which the coupling topology has had a significant effect on the dynamics but has failed to dominate it. Once again the message is not that one is more important than the other, but that a proper understanding of the system must account for the subtle interplay of both.

One further question that is of interest in a small-world context is whether or not small-world graphs are capable of phase transitions. That is, can oscillator systems that locally look like highly clustered 1-lattices display global entrainment properties similar to those of random graphs and, by association, complete graphs? To put it yet another way, starting from a one-dimensional lattice of oscillators, which synchronises only with great difficulty, can a small amount of random *rewiring* substitute for a vast amount of *additional* wiring to generate a network of otherwise identical oscillators that synchronises very easily? The answer, as Figure 9.10 shows, is "yes," on the condition that λ not be too small.[2] Figure 9.10 superposes $\gamma(\phi)$ upon $E(\phi)$ curves for $\lambda = 1$, 2, and 10. In each case the transition to global entrainment occurs whilst γ is much larger than its random limit of k/n. Recalling that a large γ combined with a small L is the hallmark of a small-world graph, then it seems that such graphs are indeed capable of phase transitions.

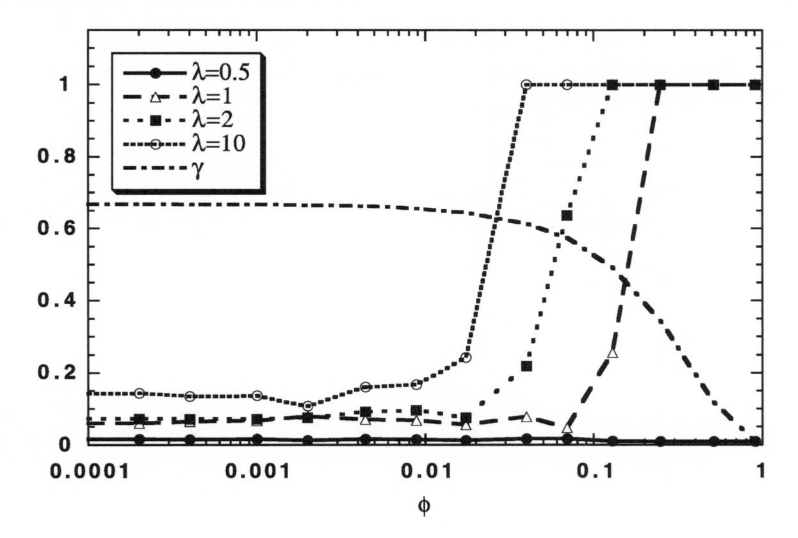

Figure 9.10 Frequency-order parameter E vs. ϕ, compared with the clustering coefficient $\gamma(\phi)$ for the underlying β-graphs ($n = 1{,}000$, $k = 1$). The transition to global entrainment occurs for $\gamma \gg k/n$ (where k/n is the expected clustering coefficient for a random graph) for all values of λ shown except $\lambda = 0.5 < \lambda_c$.

As before, it is interesting to see how things change when different graph models are used. In Chapter 8, the α- and β-graphs exhibited significantly different dynamics for the same parameter values. This does not appear to be the case here, as demonstrated in Figure 9.11. Naturally, individual random realisations vary slightly, but the qualitative behaviour is invariant with respect to the two relational-graph models. Hence all the statements above made for β-graphs can be generalised to include α-graphs as well.

Finally, what is the difference between the dynamics of relational and spatial graphs? In Chapters 3 and 4 it turned that only relational graphs were capable of small-world behaviour, as the characteristic length of spatial graphs reaches its random limit only at about the same point that the clustering coefficient reaches its random limit. Hence spatial graphs cannot exhibit both a large γ and a small L. Is this feature reflected in the dynamics of Kuramoto oscillators? Once again the answer is conditional: spatial graphs can support globally sized clusters of mutually entrained oscillators, and they can do this even when $\gamma(\xi)$ is large compared to k/n. However, this is true only for λ large compared to the equivalent λ for relational graphs. Figure 9.12 clarifies this statement.

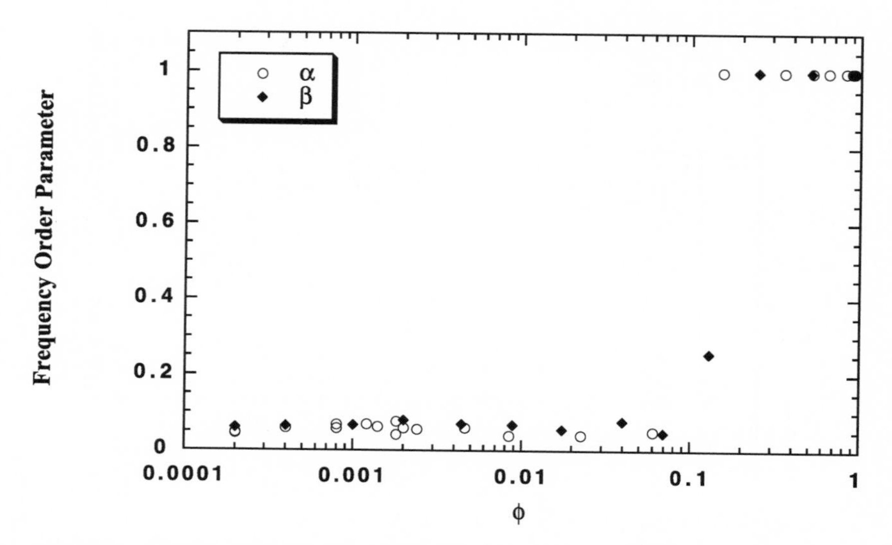

Figure 9.11 Comparison of E vs. ϕ between α- and β-graphs ($n = 1{,}000$, $k = 10$, and $\lambda = 1$).

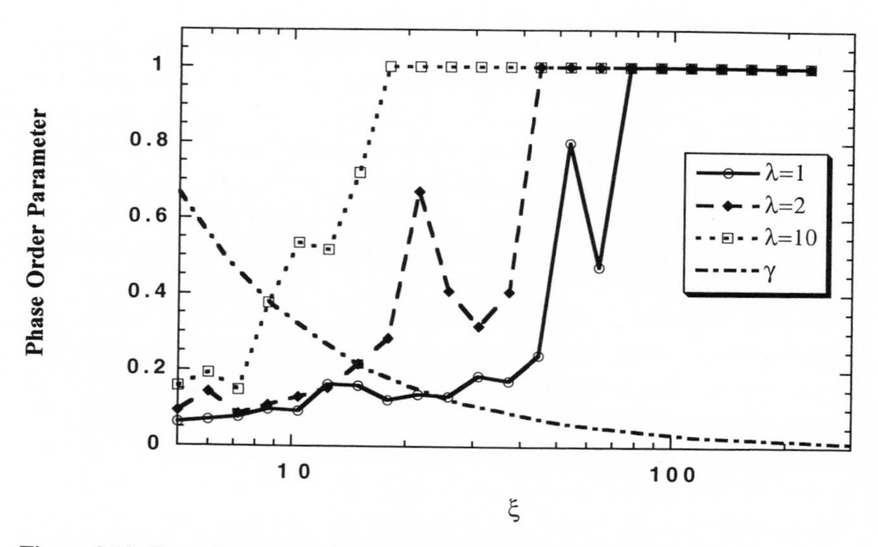

Figure 9.12 E vs. ξ compared with $\gamma(\xi)$ for uniform spatial graphs with $n = 1000$, $k = 10$. For small λ, transition to global entrainment occurs only when $\gamma \approx k/n$.

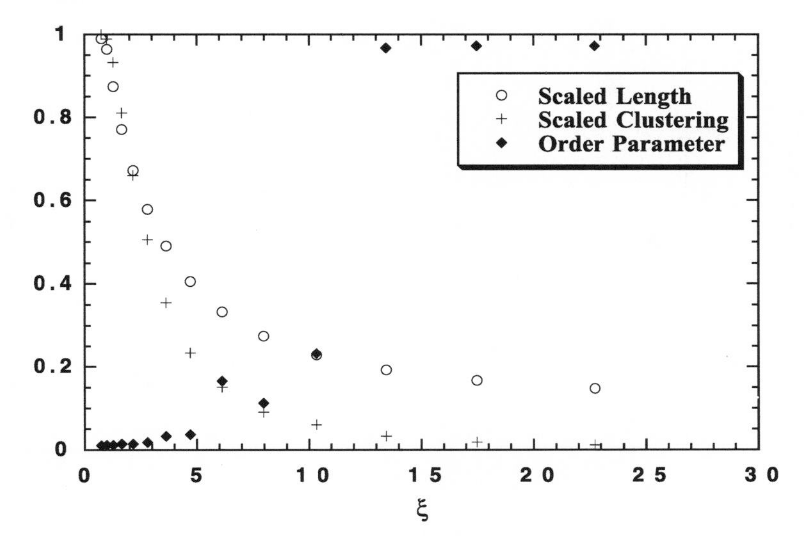

Figure 9.13 Functional comparison of L, γ, and E vs. ξ for the parameters of Nieber et al. (1991) and $\lambda = 4$. Global synchrony only occurs when $\gamma \approx k/n$.

For $\lambda = 10$, it seems clear that the system can be globally entrained and still have a relatively large γ. However, this is not true for $\lambda = 1$ and $\lambda = 2$, in which cases γ has almost reached its asymptotic limit before the respective transitions to global entrainment occur. By contrast, Figure 9.10 shows that relational graphs can support global entrainment at high γ for λ as low as $\lambda = 1$.

An interesting consequence of these results on spatial graphs is that they shed new light (or at least provide a new perspective) upon the observation of Niebur et al. mentioned earlier that long-range correlations are observed in the phases of sparsely coupled Kuramoto phase oscillators with some kind of random coupling (Niebur et al. 1991). Using a two-dimensional Gaussian spatial graph, with the parameter values of $n = 128^2 = 16{,}384$ and $k = 5$ (Niebur et al. 1991), the frequency-order parameter (E) can be computed as a function of ξ. Niebur et al. find long-range correlations at a value of the standard deviation $\sigma = 6$, which is equivalent to $\xi \approx 18$ (because, for $k = 5$, all connections are made within $\pm 3\sigma$). Figure 9.13 shows that, for $\lambda = 4$, this system does indeed achieve global frequency locking for the expected value of ξ, but not before the characteristic path length (L) and clustering coefficient (γ) have decreased to approximately their random-limit values. This is

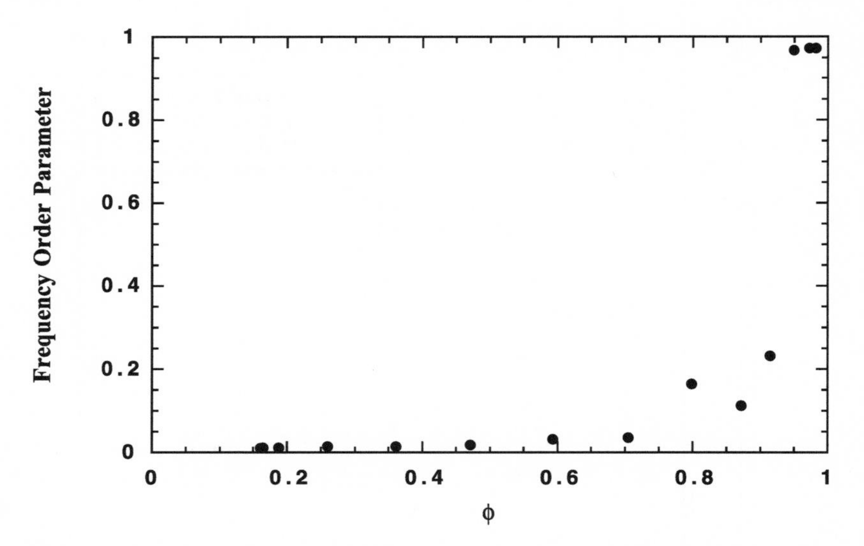

Figure 9.14 Frequency-order parameter E vs. ϕ for the parameters of Nieber et al. (1991) and $\lambda = 4$. Again, it is apparent that global synchrony only occurs in the random limit ($\phi \approx 1$).

equivalent to the statement that a Gaussian spatial graph can exhibit global frequency locking, but only if it is effectively a random graph. In this light, Niebur et al.'s result is similar to Satoh's work (1989), which showed that oscillators on random graphs were much more prone to frequency locking than on two-dimensional grids. That the graph is a random graph can be seen in a different fashion in Figure 9.14, where E is shown as a function of ϕ on the usual scale. Here it is clear that global frequency locking is only achieved in a Gaussian spatial graph, once *almost all* its edges are shortcuts. This observation is interesting because we *expect* that random graphs will exhibit superior frequency-locking capabilities, but we now know that it is possible to construct graphs that are *not* random but that can achieve a similar performance with *far fewer* shortcuts (that is, small-world graphs).

9.3 MAIN POINTS IN REVIEW

1. Kuramoto oscillators with a uniform distribution of intrinsic frequencies over the unit interval are known to exhibit a phase transition in their phase-order parameter (R) and their frequency-order parameter

(E) at a critical value of the coupling strength λ_c when they are coupled on a complete graph.

2. A similar phenomenon occurs when the oscillators are coupled according to a sparse random graph, except that the value of λ_c is shifted. Furthermore this transition appears to be invariant for large enough n, indicating that a random graph is capable of global mutual entrainment almost to the same degree as its mean-field equivalent. This result seems analogous to a random sampling problem in which the median of a distribution can be sampled to a specified precision with a sample size that is independent of the population size.

3. It also appears that this phenomenon can occur on graphs that are much less than fully random, although the corresponding phase transition occurs at increasingly large λ_c as the degree of randomness in the graph is decreased. Hence there is an interaction between the structural and dynamical parameters: for some values of ϕ a phase transition to global mutual entrainment can be generated by small changes in λ, and for some values of λ a similar transition can be generated by changing ϕ. These statements appear to be invariant with respect to α- and β-models.

4. Spatial graphs can also generate phase transitions in ξ, but a much larger λ is required than for relational graphs if the transition is to occur before ξ is so large that the graph has effectively reached its random limit.

10

Conclusions

This is not the end.
It is not even the beginning of the end.
But it is, perhaps, the end of the beginning.

Winston Churchill, 10 November 1942

The preceding chapters have attempted to justify the claim that the qualitative nature of a system's connectivity is important in determining both its structural and dynamical properties. It is important to realise, however, that this statement can be true in both a bland way and an interesting way, and that it is the interesting interpretation that has been dealt with here. In the bland version, it is obvious that topology matters, and this has been demonstrated many times over. One dimension is clearly different from two dimensions, and both are clearly different from a complete graph, or a star graph, or even a random graph. The interesting version of the "topology matters" statement, however, restricts itself to changes in topology that obey the following criteria:

1. They do not connect or disconnect the graph (all graphs are connected).
2. They do not change the average degree k.
3. They do not introduce large variations in k (that is, any more than occurs with randomly assigned edges).
4. They do not explicitly alter the dimension.

In fact, the random rewiring mechanism does not make explicit changes in topology at all, so much as it *perturbs* a single topology (the 1-lattice), keeping the total number of edges constant. Any apparent changes in topology then are really *consequences* of the random rewiring, which can be inferred by measuring the properties of the rewired graphs, not changes that were built in from the start. The graph invariants that have been defined can probably be expressed in terms of more traditional topological invariants, like dimension, and the results reinterpreted accordingly. But this is not a trivial task, and it seems not only easier but also more appropriate to leave "topological changes" in terms of the rewiring metaphor.

Even given these restrictions, it may still seem somewhat trite to say that randomly connected systems are different from systems connected like a 1-lattice. Depending on your intuition, even the occasional resemblance of random-graph dynamics to mean-field dynamics may be unsurprising, although it isn't obvious and isn't even always true (in disease-spreading, for example).

In any case, it certainly is *not* obvious that graphs with a relatively tiny fraction of shortcuts should exhibit characteristic lengths that are much closer to those of random graphs than they are to those of 1-lattices. It is even less obvious that the properties of such a wide range of dynamical systems as explored in Part II should also seem to be sensitive to small amounts of rewiring, although it should be emphasised that this is by no means a universal occurrence. The fact that small amounts of rewiring can be equivalent to large amounts of rewiring leads us to an even more striking conclusion: that small amounts of random rewiring can achieve much the same result as a massive *addition* of nonrandom edges: that is (in some instances), small-world graphs behave like random graphs that behave like complete graphs, which have vastly more edges than their sparse cousins.

A further layer of subtlety to this result (for any persistent pooh-poohers) is that not just *any* kind of random rewiring will do. This point is made clear by the comparison between relational graphs, which are not characterised by any *external* length scale, and spatial graphs, which are. This is a subtle point and one worth understanding. If shortcuts are to have the dramatic effect on length (and thus dynamics) that we see in small-world graphs, they must be able to connect vertices that would otherwise be separated by distances (graph distances, that is) on the order of the entire system. This *can* happen in spatial graphs with finite cutoffs but only if $\xi = O(n)$, in which case *most* edges will be shortcuts and the resulting clustering will be low. Conversely, if ξ is small enough such that γ will remain large, then any shortcuts must necessarily be short range. The key to generating the small-world phenomenon is the presence of a *small fraction* of very long-range, *global* edges, which contract otherwise distant parts of the graph, whilst *most* edges remain *local*, thus contributing to the high clustering coefficient. The presence of any physical length scale, *beyond which edges cannot connect*, destroys this combination and so excludes the possibility of small-world graphs.

Of course, the real story can already be seen to be more complicated than the thumbnail sketch presented here. For instance, the models constructed in Chapter 4 admit only two length scales: local and global. It is hard to imagine that reality is this forgiving, and it seems that any real problem will exhibit structure at multiple, or even a continuum of, length scales. It also seems that spatial graphs whose defining probability distributions have infinite variance will exhibit quite different properties to those that do not, but this possibility has not been investigated in detail here. Furthermore, only a handful of graph invariants have been considered. Not only are many other invariants plausible, but the current versions of those that have been defined are crude, and their purposes may be served better by more sophisticated definitions. Finally, the shortcomings in the structural analysis are beggared by the corresponding holes in the dynamical analysis. Indeed, it seems that the task of reconciling the structure and dynamics of large, sparsely coupled systems is a *big* one that will impinge upon several disciplines and arise in a multitude of applications. About all that can definitely be said at this stage is that *something* interesting is going on in a wide variety of dynamical systems. Sometimes the relevant dynamical properties seem to be dominated by the characteristic length of the graph (such as disease spreading), sometimes by the clustering (cooperation), and sometimes by both or neither (computation in CAs). Part II is more a compendium of interesting results than an analysis, and although a few explanations have been proposed for specific observations, the whole subject still seems deeply shrouded in mystery. Perhaps the main point to take away, then, is that this research illuminates an interesting and possibly important new approach to interesting and important old problems.

Notes

CHAPTER 1
KEVIN BACON, THE SMALL WORLD, AND WHY IT ALL MATTERS

1. The "Oracle of Bacon" website, created by Brett Tjaden and Glenn Wasson, is located at www.cs.virginia.edu/bct7m/bacon.html.

2. http://ww.us.imdb.com.

3. This fact, along with all others to do with the Kevin Bacon Game cited in this book, was correct as of April 1997. Since then the database has been extensively updated, but nothing essential has changed.

CHAPTER 2
AN OVERVIEW OF THE SMALL-WORLD PHENOMENON

1. This is essentially because, in a randomly connected system, the total number of members "reached" grows exponentially with increasing degree of separation, and 1,000 × 1,000 × 1,000 is greater than the population of the United States.

2. The only difference between Barnes's density and clustering is that here v is not included as a member of its own neighbourhood—an attribute of some convenience because it allows clustering of zero.

3. In the telephone book method, subjects are asked to name acquaintances who have the same last names as those that appear on an imaginary "page" of a telephone directory, which is then treated as a representative subset of the entire population.

4. This calculation works because, in a population where acquaintance probabilities are independently distributed, the number of members linked to A by an acquaintance chain of less than a specified "length" grows exponentially with the length. Hence one thousand "friends" each is many more than is required to encompass the entire U.S. population within six handshakes.

5. In other words, if a network consists of only a few elements, or if at least one element is connected to a significant fraction of the total population, then it is not surprising that it should be "small," in which case there is nothing to explain.

6. Most researchers have confined themselves to the study of connected graphs because of the obvious problems associated with the apparent infinite lengths of disconnected graphs.

7. One method relates the average distance to the independence number of the graph, that is, the maximum size of a subgraph such that every pair of vertices in the subgraph are nonadjacent. Another utilises the eigenvalues of the adjacency matrix or the closely related Laplacian matrix, which treats the graph as a system of masses coupled by linear springs in place of the edges. The eigenvalues then characterise the modes of oscillation of the resulting coupled system. See Fiedler (1973) and Cvetković et al. (1979) for an explanation of the Laplacian matrix and the properties of its eigenvalues.

8. Monotone means just that if a particular random graph G possesses Q, then any graph H that includes G as a subgraph will also have Q.

9. There are a number of technical subleties to this result. See Chapter 10 of Alon and Spencer (1992) for a reasonably accessible description of the appearance (at $k \approx 1$) of the *giant component*, which proceeds to swallow up all remaining vertices, including the last few isolates when $k \gtrsim \ln(n)$.

CHAPTER 3
BIG WORLDS AND SMALL WORLDS: MODELS OF GRAPHS

1. This rapid rise as *fraction of mutual friends* $\rightarrow 1$ is present mostly for the purpose of enforcing continuity, but it could be justified, in modelling terms, by the argument that, even in such a random world, if two people have *all* their current friends in common, then they can't really avoid knowing each other.

2. There are many reasonable choices for p, but, whilst the specific choice can affect the results quantitatively, it appears to make little *qualitative* difference so long as it is sufficiently small (that is, $p \ll \binom{n}{2}^{-1}$). Here p is set at 10^{-10} for all numerical experiments.

3. Perhaps this should not be surprising, given Bollobás's observation of (1985, p. 41) that almost all random graphs of the same order (n) and size (M) are the same (in the sense that for any property Q, almost all graphs have Q or almost none of them do), regardless of the model ($G(n, M)$ or $G(n, p)$) used to create them.

4. The data points for $L(\alpha)$ and also $\gamma(\alpha)$ were averaged over one hundred random realisations of the construction algorithm (which, recall, makes connections at random but biased by the presence of mutually adjacent vertices) to reduce statistical fluctuations. In general, these fluctuations do not affect the qualitative nature of the data. Hence for convenience (where not otherwise noted), only a single realisation of the construction algorithm is used to generate results.

5. This is simply the probability that, when a vertex creates a new edge to another vertex with uniform random probability over the entire graph, that vertex will be in the same neighbourhood.

6. Of course, for manageable values of n, $n^{1/d}$, and $\ln(n)$ can be impossible to distinguish in practice, for any but small d. Hence only two-dimensional lattice substrates will be considered in detail.

7. The main difference between a β-graph with $\beta = 1$ and a true random graph is that all vertices in the β-graph are guaranteed to have degree at least $k/2$. Vertices, however, are still *connected to* at random, so a nonzero variance in

degree still results. In effect, the increased homogeneity results in graphs whose edge distributions are representative of much larger graphs—a positive feature, as the large-n limit is the primary point of interest.

8. The issue of how small is "not too small" requires some extra work and so will be deferred until later.

9. Again, all graphs compared here have the usual parameters $n = 1{,}000$, $k = 10$ except for the the two-dimensional lattice substrate for the α-model for which $n = 1{,}024$.

10. One instance in which contractions prove indispensible is in analysing the Kevin Bacon Graph in Chapter 5.

11. More precisely, the number of edges that is rewired is $\phi kn/2$ rounded down to the nearest integer.

12. There is still a slight problem, even with this model, in that the specified ϕ may not yield an integer number of shortcuts for all n and k considered. The errors involved here are small, however, and decrease to zero in $\lim n \to \infty$.

13. Distributions with no such cutoff (and particularly those with infinite variance) raise some interesting possibilities that will be mentioned briefly at the end of Chapter 4.

14. In higher dimensions, when $\xi = \lceil k^{1/d}/2 \rceil$, then each vertex may connect with only $O(k)$ other vertices, and so an approximate d-lattice will result (exact when $d = 1$).

15. An important point to bear in mind throughout the discussion of spatial graphs is that both the distributions considered in any detail (uniform and Gaussian) exhibit a *finite cutoff*. This cutoff is exact in the case of the uniform distribution and effective in the case of the Gaussian distribution, for finite n and k. More exotic distributions with, say, infinite variance may yield qualitatively different results, but these are not considered here.

CHAPTER 4
EXPLANATIONS AND RUMINATIONS

1. For the purposes of this calculation, v is included in its own neighbourhood, and that edge is explicitly discounted.

2. A refinement of this idea would be to consider a hierarchy of length scales, but for simplicity only two will be considered here.

3. Even the definition of a contraction, based upon a single common member between groups, is a simplfied version of real networks where, in fact, large groups that would otherwise be widely separated may be joined by not one but a small number of common members. This would be a natural generalisation of the concept of contractions, but one that is beyond the scope of this treatment.

4. Actually, two vertices $u_i \in \Gamma(v)$ satisfy every value of μ, so the summation should go over this range twice. But each edge also gets counted twice (because (u_i, u_j) will get counted as an edge in both $\Gamma(u_i)$ and $\Gamma(u_j)$), and this exactly cancels the double sum.

CHAPTER 5
"IT'S A SMALL WORLD AFTER ALL": THREE REAL GRAPHS

1. See Jerry Grossman's "Erdös Number Project Home Page" (www.oakland.edu/~grossman/erdoshp.html) for some interesting facts and references.

2. A list of all authors with an Erdös Number of one is available at www.oakland.edu/~grossman/Erdos0.

3. These figures include only the cast members (no extras) of feature films, so television programmes and made-for-TV movies are not included.

4. *Express Train on a Railway Cutting* (1898), directed by Cecil M. Hepworth, was the first movie made.

5. The adjacency matrix for Kevin Bacon Graph was generously provided by Brett Tjaden, who compiled it directly from the Internet Movie Database.

6. The version of the KBG adjacency matrix used here required approximately 75 MB of ram. Hence storing a graph that was an order of magnitude larger in n would require a computer with approximately 750 MB of ram. In 1997, when the calculations were performed, a typical PC came with 32 MB of RAM, and even the Cornell supercomputer had only a handful of nodes with more than 750 MB.

7. In fact, one might even argue that Kevin Bacon has a finite Erdös Number (and thus Erdös a finite Bacon Number) because of Benedict Gross, who has Bacon Number two (via his role as a mathematical consultant in *It's My Turn*, starring Jill Clayburgh) and Erdös Number three. It is also interesting to note that the author's Ph.D. advisor (Steven Strogatz) was a student in Gross's junior seminar at Princeton in 1979 (entitled "Fermat's Last Theorem") and also got to know Alan Alda. Hence, with only a little flexing of definitions, both Strogatz and, by association, the author, have finite Bacon (and Erdös) numbers. This all seems a little beside the point, but it does serve to illustrate just how close everybody really is.

8. Bogart: *The Wagons Roll at Night* (1941); Brando: *The Teahouse of the August Moon* (1956); Burton: *The Longest Day* (1962), also starring Sean Connery and Robert Mitchum; Travolta: *The Devil's Rain* (1975); Bacon: *The Big Picture* (1989).

9. The computation of L was based upon the distribution sequence data of the KBG (provided by Brett Tjaden, Computer Science Department, University of Virginia), current as of April 1997.

10. This is one of the great advantages of having such a model, because to create $L(\psi)$ and $\gamma(\psi)$ numerically with the methods of Chapter 3 would take longer than this book took to write.

11. See Brett Tjaden's "Oracle of Bacon" web page (www.cs.virginia.edu/bct7m/bacon.html) for a list of the thousand best-connected actors and their distribution sequences.

12. As early as 1970, there were an estimated 51,000 miles of high-voltage transmission lines: a figure projected to have more than doubled by 1990 (General Electric Company, 1975, p. 13).

13. The load-flow data, upon which our adjacency matrix is based, were kindly provided by Koeunyi Bae and James Thorp, Department of Electrical Engineering, Cornell University.

14. The reason for this is obvious: k is so small that large clustering and small lengths are impossible in a connected graph.

15. In fact, the historical development of the grid supports such a conclusion. Originally the power grid was actually a number of disconnected and independent power grids, which were eventually connected for the utility of sharing surpluses and deficits and thus enhancing reliability and efficiency. The ghost of the old system remains, however, in the greater intraconnectedness of those nodes that used to belong to the same independent grids.

16. The *C. elegans* web site is at http://eatworms.swmed.edu/VLhome.shtml.

17. A one-dimensional spatial model was used as the worm is essentially extended in one-dimension, so this seemed the most appropriate as well as the simplest case to treat.

CHAPTER 6
THE SPREAD OF INFECTIOUS DISEASE IN STRUCTURED POPULATIONS

1. For other topologies, ρ_{tip} cannot be defined according to Equation 6.4, and the actual point at which a significant fraction of the population becomes infected is also far less clear.

CHAPTER 7
GLOBAL COMPUTATION IN CELLULAR AUTOMATA

1. Here k refers to the average number of neighbours per cell in the CA. To avoid confusion, it should be noted that the CA literature uses k to mean the number of states s_i (here always taken to be two) and r to refer to the radius of a neighbourhood. Hence $k = 2r$ in standard CA notation.

2. Note that this generates the "harder" binomial distribution of ρ_0 over $[0, 1]$.

3. In fact, the motivation to do the work for this section was derived from a bet with a CA specialist who suspected (having heard the idea) that small-world graphs wouldn't make any difference to majority rule's performance.

4. This reasoning is not strictly true because, whenever a cell has an equal number of on and off neighbours, it decides its state randomly. Hence the two rules cannot generate strictly opposite states. It seems, however, that over a number of time steps the random decisions balance out in each case such that the above explanation is *effectively* the case. If, instead of making a random decision every time a neighbourhood has equal numbers of on and off cells, the deciding cell in majority rules simply maintains its current state and the corresponding cell in contrarian changes, then the two rules *will* produce precisely opposite outcomes. This alternative is not adopted here, as it compromises the performance of both rules.

CHAPTER 8
COOPERATION IN A SMALL WORLD: GAMES ON GRAPHS

1. A *game* means simply any situation in which a number of entities (players) compete for limited resources according to a strict set of rules that determine, for each of a finite set of player actions, a corresponding set of payoffs. Hence a game can be something as simple as tic-tac-toe or as complicated as a model of survival behaviours evolving over multiple generations in a heterogenous ecosystem.

2. Boyd and Lorberbaum (1987) showed that TFT was not strictly evolutionarily stable, in the sense that it could be invaded by just the right mix of strategies.

3. The seed was grown by choosing a single player v at random from an entire population of initial defectors and setting its initial condition to cooperation. Its immediate neighbours, their immediate neighbours, and so on were then successively set to cooperate until the prespecified number of initial cooperators had been reached.

4. Obviously the system has only a finite number of states in its state space so it must eventually repeat. But this number is gigantic and one would have to wait around for many times the length of the age of the universe to see it. Hence *periodic orbit* means a orbit that repeats itself in considerably less than the length of time for which the system is run.

CHAPTER 9
GLOBAL SYNCHRONY IN POPULATIONS OF COUPLED PHASE OSCILLATORS

1. *No synchrony* means that no long-term synchrony occurs on a macroscopic (or global) scale. A small number of oscillators can always become aligned by random chance, hence *some* order exists ($R(t) = O(1/\sqrt{n})$) even for a completely desynchronised state.

2. This is no surprise as even the complete graph cannot exhibit a phase transition for $\lambda < \lambda_c \approx 0.64$.

Bibliography

Achacoso, T. B., and Yamamoto, W. S. (1992). *AY's Neuroanatomy of C.* elegans *for Computation.* Boca Raton, Fla.: CRC Press.

Alon, N., and Spencer, J. H. (1992). *The Probabalistic Method.* New York: Wiley.

Asimov, I. (1957). *The Naked Sun.* Garden City, N.Y.: Doubleday.

Axelrod, R. (1984). *The Evolution of Cooperation.* New York: Basic Books.

Axelrod, R., and Dion, D. (1988). The further evolution of cooperation. *Science* 242:1385–90.

Axelrod, R., and Hamilton, W. D. (1981). The evolution of cooperation. *Science* 211:1385–96.

Barabasi, A.-L., and Albert, R. (1999). Emergence of scaling in random networks. *Science* 286:509–12.

Barnes, J. A. (1969). Networks and political process. In J. C. Mitchell (ed.), *Social Networks in Urban Situations,* ch. 2, pp. 51–76. Manchester: Manchester University Press.

Barnett, G. A. (1989). Approaches to non-Euclidean network analysis. In M. Kochen (ed.), *The Small World,* ch. 17, pp. 349–72. Norwood, N.J.: Ablex.

Bernard, H. R., Johnsen, E. C., Kilworth, P. D., and Robinson, S. (1989). Estimating the size of an aveage personal network and of an event subpopulation. In M. Kochen (ed.), *The Small World,* ch. 9, pp. 159–75. Norwood, N.J.: Ablex.

Bollobás, B. (1979). *Graph Theory: An Introductory Course.* New York: Springer Verlag.

———. (1985). *Random Graphs.* London: Academic.

Bollobás, B., and Chung, F. R. K. (1988). The diameter of a cycle plus a random matching. *SIAM Journal of Discrete Mathematics* 1 (3):328–33.

Boyd, R., and Lorberbaum, J. P. (1987). No pure strategy is evolutionarily stable in the repeated prisoner's dilemma game. *Nature* 327:59.

Boyd, R., and Richerson, P. J. (1988). The evolution of reciprocity in sizable groups. *Journal of Theoretical Biology* 132:337–56.

———. (1989). The evolution of indirect reciprocity. *Social Networks* 11:213–36.

Buck, J. (1988). Synchronous rhythmic flashing of fireflies. II. *Quarterly Review of Biology* 63:265–89.

Buckley, F., and Superville, L. (1981). Distance distributions and mean distance problems. In C. C. Cadogan (ed.), *Proceedings of the Third Caribbean Conference on Combinatories and Computing,* pp. 67–76. St. Augustine: University of the West Indies.

Burks, A., ed. (1970). *Essays on Cellular Automata.* Urbana: University of Illinois Press.

Cerf, V. G., Cowan, D. D., Mullin, R. C., and Stanton, R. G. (1974). A lower bound on the average shortest path length in regular graphs. *Networks* 4:335–42.

Chowdhury, S. (1989). Optimum design of reliable IC power networks having general graph topologies. In *Proceedings of the 26th ACM/IEEE Design Automation Conference*, pp. 787–90.

Chung, F. R. K. (1986). Diameters of communication networks. In *Mathematics of Information Processing, Proceedings of Symposia in Applied Mathematics* No. 34, pp. 1–18, Providence, R.I.: American Mathematical Society.

———. (1988). The average distance and the independence number. *Journal of Graph Theory* 12 (2):229–35.

———. (1989). Diameters and eigenvalues. *Journal of the American Mathematical Society* 2 (2):187–96.

———. (1994). An upper bound on the diameter of a graph from eigenvalues associated with its Laplacian. *SIAM Journal of Discrete Mathematics* 7 (3):443–57.

Cohen, M. D., Riolo, R., and Axelrod, R. (1999). The emergence of social organisation in the Prisoner's Dilemma: how context preservation and other factors promote cooperation. Santa Fe Institute Working Paper 99-01-002.

Crutchfield, J. P. (1994). The calculi of emergence: Computation, dynamics and induction. *Physica D* 75:11–54.

Cvetković, D., Doob, M., and Sachs, H. (1979). *Spectra of Graphs: Theory and Application*. Section 8.4. New York: Academic.

Daido, H. (1988). Lower critical dimension for populations of oscillators with randomly distributed frequencies: A renormalisation group analysis. *Physical Review Letters* 61 (2):231–34.

Das, R., Crutchfield, J. P., and Mitchell, M. (1995). Evolving globally synchronized cellular automata. In L. J. Eshelman, (ed.), *Proceedings of the Sixth International Conference on Genetic Algorithms*, pp. 336–43, San Francisco: Kaufmann.

Das, R., Mitchell, M., and Crutchfield, J. P. (1994). A genetic algorithm discovers particle-based computation in cellular automata. In Y. Davidor, H. P. Schwefel, and R. Manner (eds.), *Parallel Problem Solving in Nature, Lecture Notes in Computer Science*, pp. 344–53. Berlin: Springer.

Davidson, M. L. (1983). *Multidimensional Scaling*. Wiley Series in Probability and Mathematical Statistics. New York: Wiley.

Davis, J. A. (1967). Clustering and structural balance in graphs. *Human Relations* 20:181–87.

Doriean, P. (1974). On the connectivity of social networks. *Journal of Mathematical Sociology* 3:245–58.

Doyle, J. K., and Graver, J. E. (1977). Mean distance in a graph. *Discrete Mathematics* 17:147–54.

Edelstein-Keshet, L. (1988). *Mathematical Models in Biology*. New York: Random House.

Entringer, R. C., Jackson, D. E., and Snyder, D. A. (1976). Distance in graphs. *Czechoslovak Mathematical Journal* 26:283–96.

Entringer, R. C., Meir, A., Moon, J. W., and Székely, L. A. (1994). On the Wiener index of trees from certain families. *Australasian Journal of Combinatorics* 10:211–24.

Erdös, P., and Rényi, A. (1959). On random graphs. I. *Publicationes Mathematicae (Debrecen).* 6:290–97.

———. (1960). On the evolution of random graphs. *Publication of the Mathematical Institute of the Hungarian Academy of Sciences* 5:17–61.

———. (1961a). On the evolution of random graphs. *Bulletin of the Institute of International Statistics Tokyo* 38:343–47.

———. (1961b). On the strength of connectedness of a random graph. *Acta Mathematica Scientia Hungary* 12:261–67.

Erhard, K. H., Johannes, F. M., and Dachauer, R. (1992). Topology optimization techniques for power/ground networks in VLSI. In *Proceedings of the Euro-Dac 92 European Design Automation Conference,* pp. 362–67.

Ermentrout, G. B., and Kopell, N. (1993). Inhibition-produced patterning in chains of coupled nonlinear oscillators. *SIAM Journal on Applied Mathematics* 54 (2):478–507.

Fararo, T. J., and Sunshine, M. (1964). *A Study of a Biased Friendship Net.* Syracuse, N.Y.: Syracuse University Youth Development Center and Syracuse University Press.

Felleman, D. J., and Van Essen, D. C. (1991). Distributed hierarchical processing in the primate cerebral cortex. *Cerebral Cortex* 1:1–47.

Fielder, M. (1973). Algebraic connectivity of graphs. *Czechoslovak Mathematical Journal* 23:298–305.

Foster, C. C., Rapoport, A., and Orwant, C. J. (1963). A study of a large sociogram: Elimination of free parameters. *Behavioral Science* 8:56–65.

Frank, H., and Chou, W. (1972). Topological optimization of computer networks. *Proceedings of the IEEE* 60 (11):1385–97.

Freeman, L. C., and Thompson, C. R. (1989). Estimating acquaintanceship volume. In M. Kochen, (ed.), *The Small World,* ch. 8, pp. 147–58. Norwood, N.J.: Ablex.

General Electric Company. (1975). *Transmission Line Reference Book.* Palo Alto, Calif.: Electric Power Research Institute.

Gladwell, M. (1996). The tipping point. *New Yorker* (3 June):32–38.

Glance, N. S., and Huberman, B. A. (1993). The outbreak of cooperation. *Journal of Mathematical Sociology* 17 (4):281–302.

———. (1994). The dynamics of social dilemmas. *Scientific American* (March): 76–81.

Graham, R. (1979). On properties of a well-known graph or what is your Ramsey number? In F. Harary, (ed.), *Topics in Graph Theory,* pp. 166–72. New York: New York Academy of Sciences.

Granovetter, M. S. (1973). The strength of weak ties. *American Journal of Sociology* 78 (6):1360–80.

———. (1983). The strength of weak ties: A network theory revisited. *Sociological Theory* 1:203–33.

Grossman, J. W., and Ion, P. D. F. (1995). On a portion of the well-known collaboration graph. *Congressus Numeratium* 108:129–31.

Guare, J. (1990). *Six Degrees of Separation: A Play.* New York: Vintage.

Hadley, P., Beasley, M. R., and Wiesenfeld, K. (1988). Phase locking of Josephson-junction series arrays. *Physical Review B* 38:8712–19.

Harary, F. (1959). Status and contrastatus. *Sociometry* 22:23–43.

Hassell, M. P., Comins, H. N., and May, R. M. (1994). Species coexistence and self-organizing spatial dynamics. *Nature* 370:290–92.

Herz, A. V. (1994). Collective phenomena in spatially extended evolutionary games. *Journal of Theoretical Biology* 169:65–87.

Hess, G. (1996a). Disease in metapopulation models: Implication for conservation. *Ecology* 77 (5):1617–32.

———. (1996b). Linking extinction to connectivity and habitat destruction in metapopulation models. *American Naturalist* 148 (1):226–36.

Huber, M. (1996). Estimating the average shortest path length in a graph. Technical report, Cornell University.

Huberman, B. A., and Glance, N. S. (1993). Evolutionary games and computer simulations. *Proceedings of the National Academy of Sciences* 90:7716–18.

Huygens, C. (1893). Letters to his father. In M. Nijhoff (ed.), *Oeuvrès complètes de Christian Huygens,* vol. 5, pp. 243–44. Amsterdam: Société Hollandaise des Sciences.

Longini, I. M., Jr. (1988). A mathematical model for predicting the geographic spread of new infectious agents. *Mathematical Biosciences* 90:367–83.

Kareiva, P. (1990). Population dynamics in spatially complex environments: Theory and data. *Philosophical Transactions of the Royal Society of London, Series B* 330:175–90.

Kirby, D., and Sahre, P. (1998). Six degrees of Monica. *New York Times* (21 February): Op ed. page.

Kochen, M., ed. (1989a). *The Small World.* Norwood, N.J.: Ablex.

Kochen, M. (1989b). Toward structural sociodynamics. In M. Kochen (ed.), *The Small World,* ch. 2, pp. 52–64. Norwood, N.J.: Ablex.

Kopell, N. (1988). Toward a theory of central pattern generators. In A. H. Cohen, S. Rissignol, and S. Grillner (eds.), *Neural Control of Rhythmic Movement in Vertebrates,* pp. 369–413. New York: John Wiley.

Kopell, N., and Ermentrout, G. B. (1986). Symmetry and phaselocking in chains of weakly coupled oscillators. *Communications in Pure and Applied Mathematics* 39:623–60.

Kopell, N., Zhang, W., and Ermentrout, G. B. (1990). Multiple coupling in chains of oscillators. *SIAM Journal on Mathematical Analysis* 21 (4):935–53.

Korte, C., and Milgram, S. (1970). Acquaintance linking between white and negro populations: Application of the small world problem. *Journal of Personality and Social Psychology* 15:101–18.

Kretschmar, M., and Morris, M. (1996). Measures of concurrency in networks and the spread of infectious disease. *Mathematical Biosciences* 133:165–95.

Kuramoto, Y. (1975). Self-entrainment of a population of coupled nonlinear oscillators. In H. Araki (ed.), *International Symposium on Mathematical Problems in Theoretical Physics, Lecture Notes in Physics,* vol. 39, pp. 420–22. New York: Springer.

Lin, S. (1982). Effective use of heuristic algorithms in network design. In *The Mathematics of Networks, Proceedings of Symposia in Applied Mathematics,* no. 26, pp. 63–84. Providence, R.I.: American Mathematical Society.

Lindgren, K. (1991). Evolutionary phenomena in simple dynamics. In C. G. Langton, C. Taylor, J. D. Farmer, and S. Rasmussen (eds.), *Artificial Life II, Santa Fe Institute Studies in the Sciences of Complexity*, vol. 10, pp. 295–312. New York: Addison-Wesley.

Lindgren, K., and Nordahl, M. G. (1994). Evolutionary dynamics of spatial games. *Physica D* 75:292–309.

Linial, N., London, E., and Rabinovich, Y. (1995). The geometry of graphs and some of its algorithmic applications. *Combinatorica* 15 (2):215–45.

Lorrain, F. P., and White, H. C. (1971). Structural equivalence of individuals in social networks. *Journal of Mathematical Sociology* 1:49–80.

Lumer, E. D., and Huberman, B. A. (1991). Hierarchical dynamics in large assemblies of interacting oscillators. *Physics Letters A* 160:227–32.

March, L., and Steadman, P. (1971). *The Geometry of Environment*. Chapter 14. London: RIBA Publications.

Matthews, P. C., Mirollo, R. E., and Strogatz, S. H. (1991). Dynamics of a large system of coupled nonlinear oscillators. *Physica D* 52:293–331.

May, R. M. (1995). Necessity and chance: Deterministic chaos in ecology and evolution. *Bulletin of the American Mathematical Society* 32:291–308.

May, R. M., and Nowak, M. A. (1994). Superinfection, metapopulation dynamics, and the evolution of diversity. *Journal of Theoretical Biology* 170:95–114.

Mazoyer, J. (1987). A six states minimum time solution to the firing squad synchronization problem. *Theoretical Computer Science* 50:183–238.

Milgram, S. (1967). The small world problem. *Psychology Today* 2:60–67.

———. (1969). *Obedience to Authority*. New York: Harper and Row.

Mitchell, J. C. (1969). The concept and use of social networks. In J. C. Mitchell (ed.), *Social Networks in Urban Situations*, ch. 1, pp. 1–50. Manchester: Manchester University Press.

Mitchell, M. (1996a). Computation in cellular automata. Technical report, Santa Fe Institute.

———. (1996b). *An Introduction to Genetic Algorithms. Complex Adaptive Systems.* Cambridge, Mass.: MIT Press.

Mitchell, M., Crutchfield, J. P., and Das, R. (1997). Evolving cellular automata to perform computations. In T. Back, D. Fogel, and Z. Michalewicz (eds.), *Handbook of Evolutionary Computation*. Bristol: Oxford University Press.

Mitchell, M., Crutchfield, J. P., and Hraber, P. T. (1994). Evolving cellular automata to perform computations: Mechanisms and impediments. *Physica D* 75:361–91.

Mitchell, M., Hraber, P. T., and Crutchfield, J. P. (1993). Revisiting the edge of chaos: Evolving cellular automata to perform computations. *Complex Systems* 7:89–130.

Mohar, B. (1991). Eigenvalues, diameter and mean distance in graphs. *Graphs and Combinatorics* 7:53–64.

Munkres, J. R. (1975). *Topology: A First Course*. Englewood Cliffs, N.J.: Prentice Hall.

Murray, J. D. (1993). *Mathematical Biology, 2d ed.* Berlin: Springer.

Nash, J. F. (1950). The bargaining problem. *Econometrica* 18:155–62.

———. (1951). Non-cooperative games. *Annals of Mathematics* 54:286–95.

———. (1953). Two-person cooperative games. *Ecnometrica* 21 (1):128–40.

Newman, M. E. J., Watts, D. J., and Strogatz, S. H. (2002). Random graph models of social networks. *Proceedings of the National Academy of Sciences* 99 (1):2566–72.

Niebur, E., Schuster, H. G., Kammen, D. M., and Koch, C. (1991). Oscillator-phase coupling for different two-dimensional network connectivities. *Physical Review A* 44 (10):6895–6904.

Nowak, M., and Sigmund, K. (1993). A strategy of win-stay-lose-shift that outperforms tit-for-tat in the prisoner's dilemma game. *Nature* 364:56–58.

Nowak, M. A., and Bangham, C. R. M. (1996). Population dynamics of immune responses to persistent viruses. *Science* 272:74–79.

Nowak, M. A., Bonhoeffer, S., and May, R. M. (1994). More spatial games. *International Journal of Bifurcations and Chaos* 4 (1):33–56.

Nowak, M. A., and May, R. M. (1992). Evolutionary games and spatial chaos. *Nature* 359:826–29.

———. (1993). The spatial dilemmas of evolution. *International Journal of Bifurcations and Chaos* 3 (1):35–78.

Palmer, R. (1989). Broken ergodicity. In D. L. Stein (ed.), *Lectures in the Sciences of Complexity, Santa Fe Institute Studies in the Sciences of Complexity,* no. 1, pp. 275–300. New York: Addison-Wesley.

Pippenger, N. (1982). Telephone switching networks. In *The Mathematics of Networks, Proceedings of Symposia in Applied Mathematics,* no. 26, pp. 101–33. Providence, R.I.: American Mathematical Society.

Plesnik, J. (1984). On the sum of all distances in a graph or digraph. *Journal of Graph Theory* 8:1–21.

Pollock, G. B. (1989). Evolutionary stability of reciprocity in a viscous lattice. *Social Networks* 11:175–212.

Pool, I., and Kochen, M. (1978). Contacts and influence. *Social Networks* 1:1–48.

Rapoport, A. (1953a). Spread of information through a population with socio-structural bias. I. Assumption of transitivity. *Bulletin of Mathematical Biophysics* 15:523–33.

———. (1953b). Spread of information through a population with socio-structural bias. II. Various models with partial transitivity. *Bulletin of Mathematical Biophysics* 15:535–46.

———. (1957). A contribution to the theory of random and biased nets. *Bulletin of Mathematical Biophysics* 19:257–71.

———. (1965). *Prisoner's Dilemma: A Study of Conflict and Cooperation.* Ann Arbor: University of Michigan Press.

Rouvray, D. H. (1986). Predicting chemistry from topology. *Scientific American* (September):40–47.

Russell, M. J., Switz, G. M., and Thompson, K. (1980). Olfactory influences on the human menstrual cycle. *Pharmocological Biochemical Behavior* 13:737–38.

Sakaguchi, H., Shinomoto, S., and Kuramoto, Y. (1987). Local and global self-entrainment in oscillator lattices. *Progress of Theoretical Physics* 77 (5):1005–10.

Satoh, K. (1989). Computer experiment on the cooperative behavior of a network of interacting nonlinear oscillators. *Journal of the Physical Society of Japan* 58 (6):2010–21.

Sattenspiel, L., and Simon, C. P. (1988). The spread and persistence of infectious diseases in structured populations. *Mathematical Biosciences* 90:341–66.

Schneck, R., Taylor, I., Sidman, J., and Godbole, A. (1997). Weiner index of random graphs. Technical report, personal communication.

Schuster, H. G., and Wagner, P. (1990a). A model for neuronal oscillations in the visual cortex: Mean-field theory and derivation of the phase equations. *Biological Cybernetics* 64:77–82.

———. (1990b). A model for neuronal oscillations in the visual cortex: Phase description of the feature dependent synchronization. *Biological Cybernetics* 64:83–85.

Shepard, R. N., Romney, A. K., and Nerlove, S. B. (eds.) (1972). *Multidimensional Scaling: Theory and Applications in the Behavioral Sciences.* 2 vols. New York: Seminar.

Skvoretz, J. (1985). Random and biased networks: Simulations and approximations. *Social Networks* 7:225–61.

———, and Fararo, Thomas J. (1989). Connectivity and the small world problem. In M. Kochen (ed.), *The Small World,* ch. 15, pp. 296–326. Norwood, N.J.: Ablex.

Solomonoff, R., and Rapoport, A. (1951). Connectivity of random nets. *Bulletin of Mathematical Biophysics* 13:107–17.

Satuffer, D., and Aharony, A. (1992). *Introduction to Percolation Theory.* London: Taylor and Francis.

Strogatz, S. H. (1994). Norbert Wiener's brain waves. In S. Levin (ed.), *Lecture Notes in Biomathematics,* no. 100, pp. 122–38. Berlin: Springer.

Strogatz, S. H., and Mirollo, R. E. (1988). Phase-locking and critical phenomena in lattices of coupled nonlinear oscillators with random intrinsic frequencies. *Physica D* 31:143–68.

———. (1991). Stability of incoherence in a population of coupled oscillators. *Journal of Statistical Physics* 63 (3–4):613–35.

von Neumann, J. (1966). *Theory of Self-reproducing Automata.* Urbana: University of Illinois Press.

von Neumann, J., and Morgenstern, O. (1944). *Theory of Games and Economic Behavior.* Princeton: Princeton University Press.

Wade, N. (1997). Dainty worm tells secrets of human genetic code. *New York Times* (24 June): *Science Times.*

Watts, D. J., and Strogatz, S. H. (1998). Collective dynamics of 'small-world' networks. *Nature* 393:440–42.

White, H. C. (1970). Search parameters for the small world problem. *Social Forces* 49:259–64.

White, H. C., Boorman, S. A., and Breiger, R. L. (1976). Social structure from multiple networks. I. Blockmodels of roles and positions. *American Journal of Sociology* 81 (4):730–80.

White, J. G., Southgate, E., Thompson, J. N., and Brenner, S. (1986). The structure of the nervous system of the nematode *Caenorhabditis elegans. Philosophical Transactions of the Royal Society of London, Series B* 314:1–340.

Wiener, H. (1947). Structural determination of paraffin boiling points. *Journal of the American Chemistry Society* 69:17–20.

Wilson, R. J., and Watkins, J. J. (1990). *Graphs: An Introductory Approach.* New York: Wiley.

Winfree, A. (1967). Biological rhythms and the behavior of populations of coupled oscillators. *Journal of Theoretical Biology* 16:15–42.
Winkler, P. (1990). Mean distance in a tree. *Discrete Applied Mathematics* 27:179–85.
Wolfram, S. (1983). Statistical mechanics of cellular automata. *Review of Modern Physics* 55:601–44.
———. (1984). Universality and complexity in cellular automata. *Physica D* 10:1–35.
Wolfram, S., ed. (1986). *Theory and Applications of Cellular Automata*. Singapore: World Scientific.

Index